# ESTUARIES

This volume provides researchers, students, practising engineers and managers access to state-of-the-art knowledge, practical formulae and new hypotheses for the dynamics, mixing, sediment regimes and morphological evolution in estuaries.

The objectives are to explain the underlying governing processes and synthesise these into descriptive formulae which can be used to guide the future development of any estuary. Each chapter focuses on different physical aspects of the estuarine system – identifying key research questions, outlining theoretical, modelling and observational approaches, and highlighting the essential quantitative results. This allows readers to compare and interpret different estuaries around the world, and develop monitoring and modelling strategies for short-term management issues and for longer-term problems, such as global climate change.

The book is written for researchers and students in physical oceanography and estuarine engineering, and serves as a valuable reference and source of ideas for professional research, engineering and management communities concerned with estuaries.

DAVID PRANDLE is currently Honorary Professor at the University of Wales' School of Ocean Sciences, Bangor. He graduated as a civil engineer from the University of Liverpool and studied the propagation of a tidal bore in the River Hooghly for his Ph.D. at the University of Manchester. He worked for 5 years as a consultant to Canada's National Research Council, modelling the St. Lawrence and Fraser rivers. He was then recruited to the UK's Natural Environment Research Council's Bidston Observatory to design the operational software for controlling the Thames Flood Barrier. He has subsequently carried out observational, modelling and theoretical studies of tide and storm propagation, tidal energy extraction, circulation and mixing, temperatures and water quality in shelf seas and their coastal margins.

# ESTUARIES

## Dynamics, Mixing, Sedimentation and Morphology

### DAVID PRANDLE
*University of Wales, UK*

**CAMBRIDGE**
UNIVERSITY PRESS

CAMBRIDGE UNIVERSITY PRESS
Cambridge, New York, Melbourne, Madrid, Cape Town, Singapore, São Paulo, Delhi

Cambridge University Press
The Edinburgh Building, Cambridge CB2 8RU, UK

Published in the United States of America by Cambridge University Press, New York

www.cambridge.org
Information on this title: www.cambridge.org/9780521888868

First published 2009

Printed in the United Kingdom at the University Press, Cambridge

*A catalogue record for this publication is available from the British Library*

*Library of Congress Cataloging in Publication Data*
Prandle, David.
Estuaries : dynamics, mixing, sedimentation, and morphology / David Prandle.
p.   cm.
1. Estuarine oceanography.   2. Estuaries.   I. Title.
GC97.P73   2009
551.46′18—dc22
2008019559

ISBN 978-0-521-88886-8 hardback

# Contents

# Symbols

| | |
|---|---|
| $A$ | cross-sectional area |
| $B$ | channel breadth |
| $C$ | concentration in suspension |
| $D$ | water depth |
| $E$ | vertical eddy viscosity coefficient |
| $E_X$ | tidal excursion length |
| $F$ | linearised bed friction coefficient |
| | dimensionless friction term |
| $H$ | total water depth $D + \varsigma$ |
| $I_F$ | sediment in-fill time |
| $J$ | dimensionless bed friction parameter |
| $K_z$ | vertical eddy diffusivity coefficient |
| $L$ | estuary length |
| $L_I$ | salinity intrusion length |
| $L_M$ | resonant estuarine length |
| $M_2$ | principal lunar semi-diurnal tidal constituent |
| $M_4$ | $M_6$ over-tides of $M_2$ |
| $MS_4$ | MSf over-tides of $M_2$ and $S_2$ |
| $P$ | tidal period |
| $Q$ | river flow |
| $R_I$ | Richardson number |
| $S_R$ | Strouhal number $U^*P/D$ |
| $S_c$ | Schmid number ($K_z/E$) |
| $S_t$ | Stratification number |
| $S_X$ | relative axial salinity gradient $1/\rho \; \partial\rho/\partial x$ |
| $S$ | dimensionless salinity gradient |
| SL | axial bed slope |
| SP | spacing between estuaries |

| $T_F$ | flushing time |
|---|---|
| $U$ | axial current |
| $U^*$ | tidal current amplitude |

Residual current components:

| $U_0$ | river flow |
|---|---|
| $U_s$ | density-induced |
| $U_w$ | wind-induced |
| $V$ | lateral current |
| $W$ | vertical current |
| $W_s$ | sediment fall velocity |
| $X$ | axial dimension |
| $Y$ | lateral axis |
| $Z$ | vertical axis |
| $c$ | wave celerity |
| $d$ | particle diameter |
| $f$ | bed friction coefficient ($\sim$0.0025) |
| $g$ | gravitational constant |
| $i$ | $(-1)^{1/2}$ |
| | surface slope |
| $k$ | wave number ($2\pi/\lambda$) |
| $m$ | power of axial depth variations ($x^m$) |
| $n$ | power of axial breadth variation ($x^n$) |
| $s$ | salinity |
| $t$ | time |
| $t_{50}$ | half-life of sediment in suspension ($\alpha/0.693$) |
| $y$ | dimensionless distance from mouth |
| $z$ | $= Z/D$ |
| $\alpha$ | exponential deposition rate |
| | exponential breadth variation ($e^{\alpha x}$) |
| $\tan \alpha$ | side slope gradient ($B/2D$) |
| $\beta$ | exponential suspended sediment profile |
| | exponential depth variation ($e^{\beta x}$) |
| $\gamma$ | sediment erosion coefficient |
| $\varepsilon$ | efficiency of mixing |
| $\varsigma$ | surface elevation |
| $\varsigma^*$ | tidal elevation amplitude |
| $\theta$ | phase advance of $\zeta^*$ relative to $U^*$ |
| $\lambda$ | wavelength |
| $v$ | funnelling parameter ($(n + 1)/(2 - m)$) |

| | |
|---|---|
| $\pi$ | 3.141592 |
| $\rho$ | density |
| $\sigma$ | frequency |
| $\tau$ | stress |
| $\varphi$ | latitude |
| $\varphi_E$ | potential energy anomaly |
| $\psi$ | ellipse direction |
| $\omega$ | tidal frequency ($P/2\pi$) |
| $\Omega$ | Coriolis parameter ($2\omega_s \sin\varphi$) |

*Superscripts*
   \* tidal amplitude
   − depth mean
*Subscripts*
   0 residual
   1D, 2D, 3D one-, two- and three-dimensional
Note: other notations are occasionally used locally for consistency with referenced publications. These are defined as they appear.

# 1

# Introduction

## 1.1 Objectives and scope

This book aims to provide students, researchers, practising engineers and managers access to state-of-the-art knowledge, practical formulae and new hypotheses covering dynamics, mixing, sediment regimes and morphological evolution in estuaries. Many of these new developments assume strong tidal action; hence, the emphasis is on meso- and macro-tidal estuaries (i.e. tidal amplitudes at the mouth greater than 1 m).

For students and researchers, this book provides deductive descriptions of theoretical derivations, starting from basic dynamics through to the latest research publications. For engineers and managers, specific developments are presented in the form of new formulae encapsulated within generalised Theoretical Frameworks.

Each chapter is presented in a 'stand-alone' style and ends with a concise 'Summary of Results and Guidelines for Application' outlining the issues involved, the approach, salient results and how these can be used in practical terms. The goal throughout is to explain governing processes in a generalised form and synthesise results into guideline Frameworks. These provide perspectives to interpret and inter-compare the history and conditions in any specific estuary against comparable experience elsewhere. Thus, a background can be established for developing monitoring strategies and commissioning of modelling studies to address immediate issues alongside longer-term concerns about impacts of global climate change.

### 1.1.1 Processes

Estuaries are where 'fresh' river water and saline sea water mix. They act as both sinks and sources for pollutants depending on (i) the geographical sources of the contaminants (marine, fluvial, internal and atmospheric), (ii) their biological and chemical nature and (iii) with temporal variations in tidal amplitude, river flow, seasons, winds and waves.

Tides, surges and waves are generally the major sources of energy input into estuaries. Pronounced seasonal cycles often occur in temperature, light, waves, river flows, stratification, nutrients, oxygen and plankton. These seasonal cycles alongside extreme episodic events may be extremely significant for estuarine ecology. As an example, adjustments in axial intrusion of sea water and variation in vertical stratification associated with salinity and temperature may lead to rapid colonisation or, conversely, extinction of sensitive species. Likewise, changes to the almost imperceptible larger-scale background circulations may affect the pathways and hence lead to accumulation of persistent tracers. Dyer (1997) provides further descriptions of these processes alongside useful definitions of much of the terminology used in this book.

Vertical and horizontal shear in tidal currents generate fine-scale turbulence, which determines the overall rate of mixing. However, interacting three-dimensional (3D) variations in the amplitude and phase of tidal cycles of currents and contaminants severely complicate the spatial and temporal patterns of tracer distributions and thereby the associated mixing. On neap tides, near-bed saline intrusion may enhance stability, while on springs, enhanced near-surface advection of sea water can lead to overturning. Temperature gradients may also be important; solar heating stabilises the vertical density profile, while winds promote surface cooling which can produce overturning. In highly turbid conditions, density differences associated with suspended sediment concentrations can also be important in suppressing turbulent mixing.

The spectrum of tidal energy input is effectively constrained within a few tidal constituents, and, in mid-latitudes, the lunar $M_2$ constituent is generally greater than the sum of all others – providing a convenient basis for linearisation of the equations for tidal propagation. However, 'mixing' involves a wider spectrum of interacting non-linear processes and is thus more difficult to simulate. The 'decay time' for tidal, surge, wave and associated turbulent energy in estuaries is usually measured in hours. By contrast, the flushing time for river inputs generally extends over days. Hence, simulation of the former is relatively independent of initial conditions, while simulation of the latter is complicated by 'historical' chronology resulting in accumulation of errors.

### *1.1.2 Historical developments*

Following the end of the last ice age, retreating ice cover, tectonic rebound and the related rise in mean sea level (msl) resulted in receding coastlines and consequent major changes in both the morphology and the dynamics of estuaries. Large post-glacial melt-water flows gouged deep channels with the rate of subsequent in-filling dependent on localised availability of sediments. Deforestation and subsequent changes in farming practices substantially changed the patterns of river flows and both the quantity and the nature of fluvial sediments. Thus, present-day estuarine morphologies reflect adjustments to these longer-term, larger-scale effects along-side more recent, localised impacts from urban development and engineering 'interventions'.

Ports and cities have developed on almost all major estuaries, exploiting opportunities for both inland and coastal navigation, alongside supplies of freshwater and fisheries. In more recent times, the scale of inland navigation has generally declined and the historic benefit of an estuary counterbalanced by growing threats of flooding. Since estuaries often supported major industrial development, the legacy of contaminants can threaten ecological diversity and recreational use. The spread of national and international legislation relating to water quality can severely restrict development, not least because linking discharges with resulting concentrations is invariably complicated by uncertain contributions from wider-area sources and historical residues. This combination of legal constraints and uncertainties about impacts from future climate changes threatens planning-blight for estuarine development. This highlights the need for clearer understanding of the relative sensitivity of estuaries to provide realistic perspectives on their vulnerability to change.

## 1.2 Challenges

Over the next century, rising sea levels at cities bordering estuaries may require major investment in flood protection or even relocation of strategic facilities. The immediate questions concern the changing magnitudes of tides, surges and waves. However, the underlying longer-term (decadal) issue is how estuarine bathymetries will adjust to consequent impacts on these dynamics (Fig. 1.1; Prandle, 2004). In addition to the pressing flood risk, there is growing concern about sustainable exploitation of estuaries. A common issue is how economic and natural environment interests can be reconciled in the face of increasingly larger-scale developments.

### *1.2.1 Evolving science and technology agendas*

Before computers became available, hydraulic scale models were widely used to simulate dynamics and mixing in estuaries. The scaling principles were based on maintaining the ratios of the leading terms in the equation describing tidal propagation. Ensuing model 'validation' was generally limited to reproduction of tidal heights along the estuary. Subsequent expansion in observational capabilities indicated how difficulties arose when such models were used to study saline intrusion, sediment regimes and morphological adjustments.

Even today, validation of sophisticated 3D numerical models may be restricted to simulation of an $M_2$ cycle – providing little guarantee of accurate reproduction of higher harmonics or residual features. Likewise, these fine-resolution 3D models

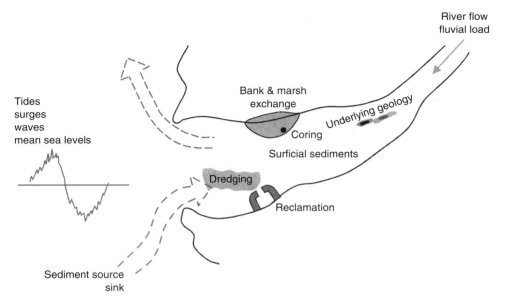

Fig. 1.1. Schematic of major factors influencing estuarine bathymetry.

may encounter difficulties in reproducing the complexity and diversity of mixing and sedimentary processes. Moreover, the paucity of observational data invariably limits interpretation of sensitivity tests. However, modelling is relatively cheap and continues to advance rapidly, whilst observations are expensive and technology developments often take decades. Thus, a major challenge in any estuary study is how to use theory to bridge the gaps between modelling and available observations. Both historical and 'proxy' data must be exploited, e.g. wave data constructed from wind records, flood statistics from adjacent locations, sedimentary records of flora and fauna as indicators of saline intrusion and anomalous fossilised bed features as evidence of extreme events.

The evolving foci for estuarine research are summarised in Fig. 1.2. These have evolved alongside successive advances in theory, modelling and observational technologies to address changing political agendas.

### *1.2.2 Key questions*

Successive chapters address the following sequence of key questions:

(Q1)  How can strategies for sustainable exploitation of estuaries be developed?
(Q2)  How do tides in estuaries respond to shape, length, friction and river flow? Why are some tidal constituents amplified yet others reduced and why does this vary from one estuary to another?

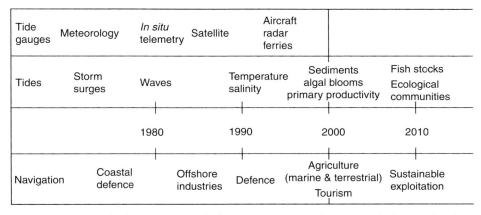

| Tide gauges | Meteorology | *In situ* telemetry | Satellite | Aircraft radar ferries | |
|---|---|---|---|---|---|
| Tides | Storm surges | Waves | Temperature salinity | Sediments algal blooms primary productivity | Fish stocks Ecological communities |
| | 1980 | | 1990 | 2000 | 2010 |
| Navigation | Coastal defence | Offshore industries | Defence | Agriculture (marine & terrestrial) Tourism | Sustainable exploitation |

Fig. 1.2. Historical development in key processes, 'end-users' and observational technologies.

(Q3) How do tidal currents vary with depth, friction, latitude and tidal period?

(Q4) How does salt water intrude and mix and how does this change over the cycles of Spring–Neap tides and flood-to-drought river flows?

(Q5) How are the spectra of suspended sediments determined by estuarine dynamics?

(Q6) What determines estuarine shape, length and depth?

(Q7) What causes trapping, sorting and high concentrations of suspended sediments? How does the balance of ebb and flood sediment fluxes adjust to maintain bathymetric stability?

(Q8) How will estuaries adapt to Global Climate Change?

## 1.3 Contents

### *1.3.1 Sequence*

The chapters follow a deductive sequence describing (2) Tidal Dynamics, (3) Currents, (4) Saline Intrusion, (5) Sediment Regimes, (6) Synchronous Estuary: Dynamics, Saline Intrusion and Bathymetry, (7) Synchronous Estuary: Sediment Trapping and Sorting – Stable Morphology and (8) Strategies for Sustainability. Analytical solutions for the first-order dynamics of estuaries are derived in Chapter 2 and provide the basic framework of our understanding. Details of associated currents are described in Chapter 3. Tidal currents and elevations in estuaries are largely independent of biological, chemical and sedimentary processes – except for their influences on the bed friction coefficient. Conversely, these latter processes are generally highly dependent on tidal motions. Thus, in Chapters 4 and 5, we consider how estuarine mixing and sedimentation are influenced by tidal action. Chapters 6 and 7 apply these theories to synchronous estuaries, yielding explicit algorithms for tidal currents, estuarine lengths and depths, sediment sorting and trapping and a bathymetric framework based on tidal amplitude and river flow.

### *1.3.2 Tidal dynamics*

Chapter 2 examines the propagation of tides, generated in ocean basins, into estuaries, explaining how and why tidal elevations and currents vary within estuaries (Fig. 1.3; Prandle, 2004). The mechanisms by which semi-diurnal and diurnal constituents of ocean tides produce additional higher-harmonic and residual components within estuaries are illustrated. Since the expedient of linearising the relevant equations in terms of a predominant ($M_2$) constituent is extensively used throughout this book, the details of this process are described. Many earlier texts and much of the literature focus on large, deep estuaries with relatively low friction effects. Here, it is indicated how to differentiate between such deep estuaries and shallower frictionally dominated systems and the vast differences in their response characteristics are illustrated.

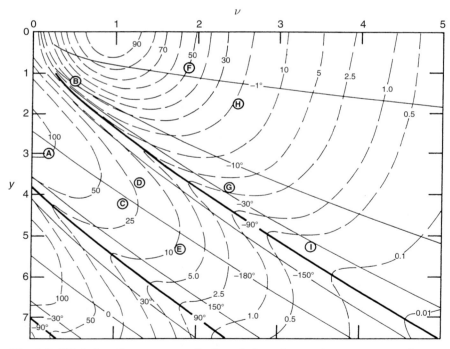

Fig. 1.3. Tidal elevation responses for funnel-shaped estuaries. $\nu$ represents degree of bathymetric funnelling and $y$ distance from the mouth, $y = 0$. Dashed contours indicate relative amplitudes and continuous contours relative phases. Lengths, $y$ (for $M_2$), and shapes, $\nu$, for estuaries (A)–(I) shown in Table 2.1.

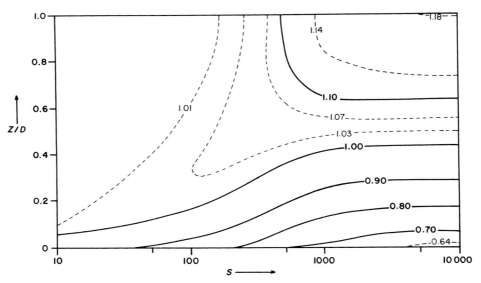

Fig. 1.4. Vertical profiles of tidal current, $U^*(z)/U^*_{mean}$, versus the Strouhal number, $S_R$, $U^*$ tidal current amplitude, $P$ tidal period, $D$ depth, $S_R = U^*P/D$.

### *1.3.3  Currents*

Chapter 3 examines how tidal currents vary along (axially) and across estuaries and from surface to bed. Changes in current speed, direction and phase (timing of peak or slack values) are explained by decomposition of the tidal current ellipse into clockwise and anti-clockwise rotating components. While the main focus is on explaining the nature and range of tidal currents, the characteristics of wind- and density-driven currents are also described. A particular emphasis is on deriving the scaling factors which encapsulate the influence of the ambient environmental parameters, namely depth, friction factor and Coriolis coefficient, i.e. latitude (Fig. 1.4; Prandle, 1982).

### *1.3.4  Saline intrusion*

Noting the earlier definition of estuaries as regions where salt and fresh water mix, Chapter 4 examines the details of this mixing. It is shown how existing theories derived for saline intrusion in channels of constant cross section can be adapted for mixing in funnel-shaped estuaries. Saline intrusion undergoes simultaneous adjustments in axial location and mixing length – explaining traditional problems in understanding observed variations over spring–neap and flood-drought conditions (Fig. 1.5; Liu *et al.*, 2008).

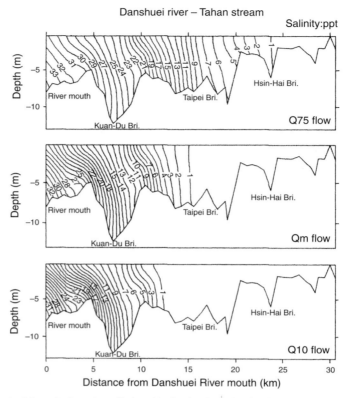

Fig. 1.5. Axial variations in salinity, ‰, in the Danshuei River, Taiwan Q75, flow rate exceeded 75% of time, Q10 flow exceeded 10% of time.

The predominance of mixing by vertical stirring driven by tidally induced turbulence has long been recognised. Here, the importance of incorporating the effects of tidal straining and resultant convective overturning is described.

The ratio of currents, $U_0/U^*$, associated with river flow and tides, is shown to be the most direct determinant of stratification in estuaries.

### 1.3.5  Sediment regimes

Chapter 5 focuses on the character of sediment regimes in strongly tidal estuaries, adopting a radically different approach to traditional studies of sediment regimes.

Analytical solutions are derived encapsulating and integrating the processes of erosion, suspension and deposition to provide descriptions of the magnitude, time series and vertical structure of sediment concentrations. These descriptions enable the complete range of sediment regimes to be characterised in terms of varying sediment type, tidal current speed and water depth (Fig. 1.6; Prandle, 2004). Theories are

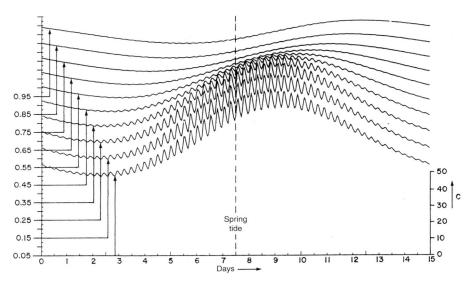

Fig. 1.6. Spring–neap patterns of sediment concentrations at fractional heights above the bed.

developed by which tidal analyses of suspended sediment time series, obtained from either model simulations or observations, can be used to explain the underlying characteristics.

### 1.3.6 Synchronous estuary: dynamics, saline intrusion and bathymetry

A 'synchronous estuary' is where the sea surface slope due to the axial gradient in phase of tidal elevation significantly exceeds the gradient from changes in tidal amplitude. The adoption of this assumption in Chapters 6 and 7 enables the theoretical developments described in earlier chapters to be integrated into an analytical emulator, incorporating tidal dynamics, saline intrusion and sediment mechanics. Chapter 6 re-examines the tidal response characteristics for any specific location within an estuary. The 'synchronous' assumption yields explicit expressions for both the amplitude and phase of tidal currents and the slope of the sea bed. Integration of the latter expression provides an estimate of the shape and length of an estuary. By combining these results with existing expressions for the length of saline intrusion and further assuming that mixing occurs close to the seaward limit, an expression linking depth at the mouth with river flow is derived. Hence, a framework for estuarine bathymetry is formulated showing how size and shape are determined by the 'boundary conditions' of tidal amplitude and river flow (Fig. 1.7; Prandle *et al.*, 2005).

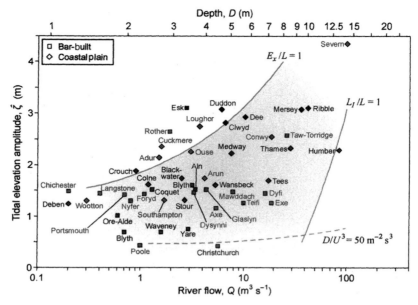

Fig. 1.7. Zone of estuarine bathymetry. Coordinates $(Q, \varsigma)$ for Coastal Plain and Bar-Built estuaries, $Q$ river Flow and $\varsigma$ elevation amplitude. Bathymetric zone bounded by $E_x < L$, $L_1 < L$ and $D/U^3 < 50\,\mathrm{m^2\,s^{-3}}$.

### *1.3.7 Synchronous estuary: sediment trapping and sorting – stable morphology*

Chapter 7 indicates how, in 'synchronous' estuaries, bathymetric stability is maintained via a combination of tidal dynamics and 'delayed' settlement of sediments in suspension. An analytical emulator integrates explicit formulations for tidal and residual current structures together with sediment erosion, suspension and deposition. The emulator provides estimates of suspended concentrations and net sediment fluxes and indicates the nature of their functional dependencies. Scaling analyses reveal the relative impacts of terms related to tidal non-linearities, gravitational circulation and 'delayed' settling.

The emulator is used to derive conditions necessary to maintain zero net flux of sediments, i.e. bathymetric stability. Thus, it is shown how finer sediments are imported and coarser ones are exported, with more imports on spring tides than on neaps, i.e. selective trapping and sorting and consequent formation of a turbidity maximum. The conditions derived for maintaining stable bathymetry extend earlier concepts of flood- and ebb-dominated regimes. Interestingly, these derived conditions correspond with maximum sediment suspensions. Moreover, the associated sediment-fall velocities are in close agreement with settling rates observed in many estuaries. Figure 1.8 (Lane and Prandle, 2006) encapsulates these results, illustrating the dependency on

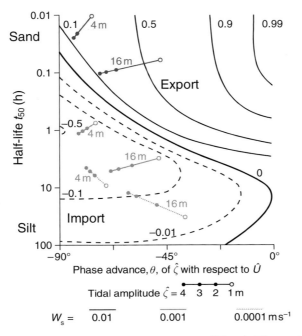

Fig. 1.8. Net import versus export of sediments as a $f(\theta, t_{50})$. Theoretical contours from (7.33). Specific examples of spring–neap variability for tidal amplitudes $\varsigma = 1$ (open circle), 2, 3 and 4 m; fall velocities, $W_s = 0.0001$, 0.001 and 0.01 m s$^{-1}$ and depths, $D = 4$ and 16 m.

delayed settlement (characterised by the half-life in suspension $t_{50}$) and the phase difference, $\theta$, between tidal current and elevation. A feedback mechanism between tidal dynamics and net sedimentation/erosion is identified involving an interaction between suspended and deposited sediments.

These results from Chapters 6 and 7 are compared with observed bathymetric and sedimentary conditions over a range of estuaries in the USA, UK and Europe. By encapsulating the results in typological frameworks, the characteristics of any specific estuary can be immediately compared against these theories and in a perspective of other estuaries. Identification of 'anomalous' estuaries can provide insight into 'peculiar' conditions and highlight possible enhanced sensitivity to change. Discrepancies between observed and theoretical estuarine depths can be used to estimate the 'age' of estuaries based on the intervening rates of sea level rise.

Importantly, the new dynamical theories for estuarine bathymetry take no account of the sediment regimes in estuaries. Hence, the success of these theories provokes a reversal of the customary assumption that bathymetries are determined by their prevailing sediment regimes. Conversely, it is suggested that the prevailing sediment regimes are in fact the consequence of rather than the determinant for estuarine bathymetries.

### *1.3.8 Strategies for sustainability*

Global climate change threatens to increase the risk of flooding in estuaries world-wide. To address this threat and to maintain a balance between exploitation and conservation, there is an urgent need for improved scientific understanding, expressed in computer-based models that are able to differentiate and predict the impact of human's activities from natural variability. Long-term data sets are vital for such understanding. Systematic marine-monitoring programmes are required, involving combinations of remote sensing, moorings and coastal stations. Likewise, continued development of Theoretical Frameworks is necessary to interpret ensemble modelling sensitivity simulations and to reconcile disparate findings from the diverse range of estuarine types.

In Chapter 8, developments in modelling, observational technologies and theory are reviewed with a detailed study of the Mersey Estuary used as a test case. Using

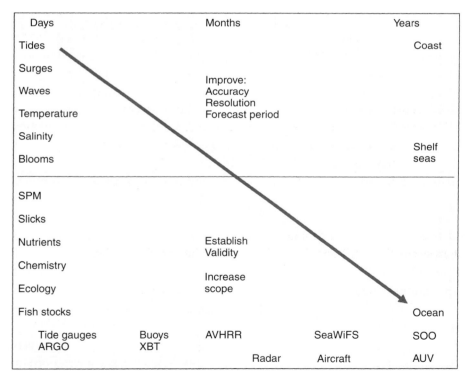

Fig. 1.9. Model evolution: extending parameters, observational technologies, time and space scales.

the theories developed in earlier chapters, estimates of likely impacts of global climate change are quantified across a range of estuaries. It is emphasised how international co-operation is necessary to access the resources required to ameliorate the threats to the future viability of estuaries.

## 1.4 Modelling and observations

Since this book focuses on the development of theories for underpinning modelling and planning measurements, a background to the capabilities and limitations of models and observations is presented.

### *1.4.1 Modelling*

Models synthesise theory into algorithms and use observations to set-up, initialise, force, assimilate and evaluate simulations in operational, pre-operational and 'exploratory' modes (Appendix 8A). The validity of models is limited by the degree to which the equations or algorithms synthesise the governing processes and by numerical and discretisation accuracies. The accuracy of model simulations depends further on the availability and suitability (accuracy, resolution, representativeness and duration) of data from observations and linked models (adjacent sea, meteorological and hydrological).

Parameters of interest include tides, surges, waves, currents, temperature, salinity, turbidity, ice, sediment transport and an ever-expanding range of biological and chemical components. The full scope of model simulations spans across atmosphere–seas–coasts–estuaries, between physics–chemistry–biology–geology–hydrology and extends over hours to centuries and even millennia. Recent developments expand to total-system simulators embedding the models described here within socio-economic planning scenarios.

### *Resolution*

Models can be (i) non-dimensional conceptual modules encapsulated into whole-system simulations, (ii) one-dimensional (1D), single-point vertical process studies or cross-sectionally averaged axial representations, (iii) two-dimensional (2D), vertically averaged representations of horizontal circulation or (iv) fully 3D. Over the past 40 years, numerical modelling has developed rapidly in scope, from hydrodynamics to ecology, and in resolution, progressing from the earliest 1D barotropic models to present-day 3D baroclinic – incorporating evolving temperature- and salinity-induced density variations. Comparable resolutions have expanded from typically 100 axial sections to millions of elements, exploiting the contemporaneous development of computing power. Unfortunately, concurrent development in observational capabilities has not kept pace, despite exciting advances in areas such as remote sensing and sensor technologies.

Tidal predictions for sea level at the mouth of estuaries have been available for more than a century. The dynamics of tidal propagation are almost entirely determined by a combination of tides at the mouth and estuarine bathymetry with some modulation by bed roughness and river flows. Thus, 1D models, available since the 1960s, can provide accurate simulation of the propagation of tidal heights and phases. However, tidal currents vary over much shorter spatial scales reflecting localised changes in bathymetry, creating small-scale variability in both the vertical and the horizontal dimensions. Continuous growth in computer power has enabled these 1D models to be extended to two and three dimensions, providing the resolution necessary to incorporate such variability. The full influence of turbulence on the dynamics of currents and waves and their interaction with near-bed processes remains to be clearly understood. Presently, most 3D estuarine models use a 1D (vertical) turbulence module. Development of turbulence models is supported by new measuring techniques like microstructure profilers which provide direct comparisons with simulated energy dissipation rates.

These latest models can accurately predict the immediate impact on tidal elevations and currents of changes in bathymetry (following dredging or reclamation), river flow or bed roughness (linked to surficial sediments or flora and fauna). Likewise, such models can provide estimates of the variations in salinity distributions (ebb to flood, spring to neap tides, flood to drought river flows), though with a reduced level of accuracy. The further step of predicting longer-term sediment redistributions remains problematic. Against a background of subtly changing chemical and biological mediation of estuarine environments, specific difficulties arise in prescribing available sources of sediment, rates of erosion and deposition, the dynamics of suspension and interactions between mixed sediment types.

Higher resolution can provide immediate improvements in the accuracy of simulations. Similarly, adaptable and flexible grids alongside more sophisticated numerical methods can reduce problems of 'numerical dispersion'. In the horizontal, rectangular grids are widely used, often employing polar coordinates of latitude and longitude. Irregular grids, generally triangular or curvi-linear, are used for variable resolution. The vertical resolution may be adjusted for detailed descriptions – near the bed, near the surface or at the thermocline. The widely used sigma coordinate system accommodates bottom-following by making the vertical grid size proportional to depth. In computational fluid dynamics, continuously adaptive grids provide a wide spectrum of temporal and spatial resolution especially useful in multi-phase processes.

Broadly, first-order dynamics are now well understood and can be accurately modelled. Hence, research focuses on 'second-order' effects, namely higher-order (and residual) tides; vertical, lateral and high-frequency variability in currents; and salinity. For the pressing problems concerning the net exchange of contaminants, non-linear interactions are important, and accurate time-averaging requires second-order accuracy in the temporal and spatial distributions of currents, elevations and density. Numerical simulation of these higher-order effects requires increasingly fine resolution. Thus, ironically, despite the exponential growth in computer power since the first 1D tidal models, limitations in computing power remain an obstacle to progress.

### *1.4.2 Observations*

Rigorous model evaluation and effective assimilation of observational data into models require broad compatibility in their respective resolution and accuracy – temporally and spatially across the complete parameter range. Technologies involved in providing observational data range from development of sensors and platforms, design of optimal monitoring strategies to analyses, curation and assimilation of data.

The set-up of estuarine models requires accurate fine-resolution bathymetry, and ideally, corresponding descriptions of surficial sediments/bed roughness. Subsequent forcing requires tide, surge and wave data at the open-sea boundary together with river flows at the head alongside their associated temperature, sediment and ecological signatures.

Sensors use mechanical, electromagnetic, optical and acoustic media. Platforms extend from *in situ*, coastal, vessel-mounted to remote sensing, e.g. satellites, aircraft, radar, buoys, floats, moorings, gliders, automated underwater vehicles (AUVs), instrumented ferries and shore-based tide gauges.

Remote-sensing techniques have matured to provide useful descriptions of ocean wind, waves, temperature, ice conditions, suspended sediments, chlorophyll, eddy and frontal locations. Unfortunately, these techniques provide only sea-surface values, and *in situ* observations are necessary both for vertical profiles and to correct for atmospheric distortion in calibration. The improved spatial resolution provided from aircraft surveillance is especially valuable in estuaries. High-frequency radars also provide synoptic surface fields of currents, waves and winds on scales appropriate to the validation of estuarine models.

It is convenient to regard observational programmes in three categories: measurements, observations and monitoring. Process measurements aim to understand specific detailed mechanics, often with a localised focus over a short period, e.g. derivation of an erosion formula for extreme combinations of tides and waves. Test-bed observations aim to describe a wide range of parameters over a wide area over a prolonged period (spanning the major cycles of variability). Thus, year-long measurements of tides, salinity and sediment distributions throughout an estuary provide an excellent basis for calibrating, assessing and developing a numerical modelling programme. Monitoring implies permanent recording, such as tide gauges. Careful site selection, continuous maintenance and sampling frequencies sufficient to resolve significant cycles of variability are essential. A comprehensive monitoring strategy is likely to embed all three of the above and include duplication and synergy to address quality assurance issues. Models can be used to identify spatial and temporal modes and scales of coherence to establish sampling resolution and to optimise the selection of sensors, instruments, platforms and locations. Coastal observatories now extend observational programmes to include physical, chemical and biological parameters.

### *Teleconnections*

In addition to the immediate, localised requirements, information may be needed about possible changes in ocean circulation which may influence regional climates and the supplies and sinks for nutrients, contaminants, thermal energy, etc. Associated data are provided by meteorological, hydrological and shelf-sea models. Ultimately, fully coupled, real-time (operational) global models will emerge incorporating the total water cycle (Appendix 8A). The large depths of the oceans introduce long inertial lags in impacts from Global Climate Change. By contrast, in shallow estuaries, detection of systematic regional variations may provide early warning of impending impacts.

### 1.5 Summary of formulae and theoretical frameworks

The following lists summarise formulae and Theoretical Frameworks presented in following chapters.

| Parameter | Dependencies | | Equation number |
|---|---|---|---|
| (a) Current amplitude | $U* \propto \varsigma^{1/2} D^{1/4} f^{-1/2}$ | shallow water | (6.9) |
|  | $\propto \varsigma D^{-1/2}$ | deep water | (6.9) |
| (b) Estuarine length | $L \propto D^{5/4}/\varsigma^{1/2} f^{1/2}$ | | (6.12) |
| (c) Depth at the mouth | $D_0 \propto (\tan \alpha \, Q)^{0.4}$ | | (6.25) |
| (d) Depth variation | $D(x) \propto D_0 x^{0.8}$ | | (6.11) |
| (e) Ratio of friction: inertia | $F/\omega \propto 10 \, \varsigma /D$ | | (6.8) |
| (f) Stratification limit | $\varsigma \sim 1 \, \text{m}$ | | (6.24) |
| (g) Salinity intrusion | $L_I, \propto D^2/f \, U_0 \, U*$ | | (6.16) |
| (h) Bathymetric zone | $L_I < L, \, E_X < L$ and $D/U*^3$ $< 50 \, \text{m}^{-2} \text{s}^3$ | | (6.23) |
| (i) Flushing time | $T_F \propto L_I/U_0$ | | (6.17) |
| (j) Suspended concentration | $C \propto f \, U*$ | | (7.36) |
| (k) Equilibrium fall velocity | $W_s \propto f \, U*$ | | (7.35) |

$\varsigma$ is tidal elevation amplitude, $f$ bed friction coefficient, $Q$ river flow with current speed $U_0$, $\tan \alpha$ lateral inter-tidal slope, $F$ linearised friction coefficient, $\omega$ tidal frequency, $E_x$ tidal excursion.

Theoretical frameworks have been established to explain both amplitude and phase variations of elevations and (cross-sectionally averaged) currents for the primary tidal constituents. Qualitative descriptions of vertical current structure have been derived for (i) oscillatory tidal components and (ii) residual components associated with river flow, wind forcing and both well-mixed and fully stratified density gradients. These dynamical results provide the basis for similar frameworks describing saline intrusion and sedimentation. Further applications of these theories for synchronous estuaries enable the frameworks to be extended to illustrate conditions corresponding to stable bathymetry and sedimentary regimes.

| Theoretical Framework | Figure | Question |
|---|---|---|
| (**T1**) Tidal response | 2.5 | **Q2** |
| (**T2**) Current structure: (a) tidal | 3.3 | **Q3** |
| (b) riverine, wind and density gradient | 4.4 | |
| (**T3**) Saline mixing | 4.13 | |
| (**T4**) Sediment concentrations | 5.6 | **Q5** |
| (**T5**) Bathymetry: | | |
| (a) Bathymetric zone | 6.12 | **Q6** |
| (b) Stability | 7.7 | **Q7** |
| (c) Lengths and depths | 8.7 | |
| (d) $W_s$, $C$, $T_F$ | 7.11 | |

$W_s$ is the fall velocity for stable bathymetry, $C$ is mean suspended sediment concentration and $T_F$ is flushing time. Q2 to Q8 refer to cardinal questions highlighted in the Summary Sections of Chapters 2 to 8. Equation (4.44) addresses Q4 and Figs 7.9 and 7.10 and Table 8.4 address Q8. Figure 1.2 and Section 1.4 summarise the issues concerned in Q1.

# Appendix 1A

## *1A.1  Tide generation*

Much of the theory presented here focuses on strongly tidal estuaries where the $M_2$ constituent amplitudes are used as a basis for parameterising the linearised bed-friction coefficient, eddy viscosity and diffusivity together with related half-lives of sediments in suspension. Figure 1A.1 shows tidal elevations in the Mersey, illustrating the predominance of the semi-diurnal $M_2$ constituent. Here, we introduce a brief background to the generation of tides, illustrating their spectral and latitudinal variations. For a rigorous, historical account of the development of tidal theory see

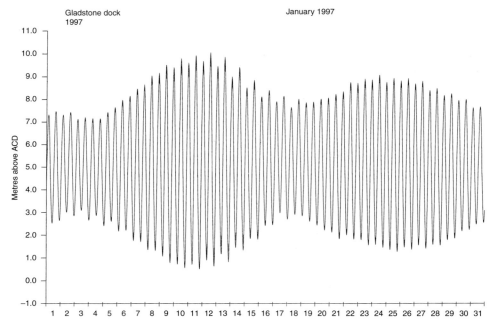

Fig. 1A.1. Month-long recording of tidal heights at the mouth of the Mersey Estuary.

Cartwright (1999). For pictorial illustrations and simplified deductive steps for the following theory see Dean (1966).

Newton's gravitational theory showed that the attractive force between bodies is proportional to the product of their mass divided by the square of their distance apart. This means that only the tidal effects of the Sun and the Moon need be considered. Mathematically, it is convenient to regard the Sun as rotating around a 'fixed' Earth – enabling the same theory to be applied to the attraction from both the Sun and the Moon.

## 1A.2 Non-rotating Earth

The attractive force on the Earth's surface due to the Moon's orbit can be separated into two components:

$$\text{tangential} \qquad \frac{3}{2} g \frac{M}{E} \left(\frac{a}{d}\right)^3 \sin 2\theta \qquad\qquad (1A.1)$$

$$\text{radial} \qquad g \frac{M}{E} \left(\frac{a}{d}\right)^3 (1 - 3\cos^2\theta), \qquad\qquad (1A.2)$$

where $M/E$ is the ratio of the mass of the Moon to that of the Earth, i.e. $1/81$, and $a/d$ is the ratio of the radius of the Earth to their distance apart, i.e. $1/60$. The longitude,

$\theta$, is measured relative to their alignment along the ecliptic plane of the Moon's orbit. The radial force component is negligible compared with gravity, $g$.

Integrating the tangential force, with the constant of integration determined from satisfying mass conservation, indicates a surface displacement:

$$\eta = \frac{a\,M}{4\,E}\left(\frac{a}{d}\right)^{3}(3\cos 2\theta + 1). \tag{1A.3}$$

This corresponds to bulges on the sides of the Earth nearest and furthest from the Moon of about 35 cm, with depressions at the 'poles' of about 17 cm.

### 1A.3 Rotating Earth

Taking account of the Earth's rotation, $\cos\theta = \cos\phi \cdot \cos\lambda$, where $\phi$ is latitude and $\lambda$ the angular displacement and hence

$$\eta = \frac{a\,M}{2\,E}\left(\frac{a}{d}\right)^{3}(3\,\cos^{2}\phi\,\cos^{2}\lambda - 1). \tag{1A.4}$$

Thus, we note the generation of two tides per day (semi-diurnal) with maximum amplitude at the equator $\phi = 0$ and zero at the poles $\phi = 90°$. The period of the principal solar semi-diurnal constituent, $S_2$, is 12.00 h. The Moon rotates in 27.3 days, extending the period of the principal lunar semi-diurnal constituent to 12.42 h. The ubiquitous spring–neap variations in tides follow from successive intervals of coincidence and opposition of the phases of $M_2$ and $S_2$. The two constituents are in phase when the Sun and the Moon are aligned with the Earth, i.e. both at 'full moon' and 'new moon'.

### 1A.4 Declination

The Moon's orbit is inclined at about 5° to the equator; this introduces a daily inequality in (1A.4), producing a principal lunar diurnal constituent, $O_1$. The equivalent solar declination is 27.3°, producing the principal solar diurnal constituent $P_1$ alongside the principal lunar and solar constituent $K_1$. The lunar declination varies over a period of 18.6 years changing the magnitude of the lunar constituents by up to $\pm 4\%$.

### 1A.5 Elliptic orbit

The Moon and the Sun's orbits show slight ellipticity, changing the distance $d$ in (1A.4). For the Moon, this introduces a lunar ellipse constituent $N_2$, while for the Sun constituents at annual, Sa, and semi-annual period, Ssa, are introduced.

### 1A.6  Relative magnitude of the Sun's attraction

Although the ratio of masses, $S/E = 3.3 \times 10^5$, overshadows that of $M/E$, this is counterbalanced by the corresponding ratio of distances $d_s/d_m \sim 390$. Thus, the relative impact of Moon: Sun is given from (1A.4) as $(S/M)/(d_s/d_m)^3 \sim 0.46$.

### 1A.7  Equilibrium constituents

In consequence of the above, 'equilibrium' magnitudes of the principal constituents relative to $M_2$ are $S_2 -0.46$, $N_2 -0.19$, $O_1 -0.42$, $P_1 -0.19$ and $K_1 -0.58$.

### 1A.8  Tidal amphidromes

The integration of tidal potential over the spatial extent of the deep oceans means that 'direct' attraction in adjacent shelf seas can be neglected compared with the propagation of energy from the oceans. In consequence, tides in enclosed seas and lakes tend to be minimal. In practice, the world's oceans respond dynamically to the above tidal forces. Responses in ocean basins and within shelf seas take the form of amphidromic systems – as shown in Fig. 1A.2 (Flather, 1976) for the $M_2$ constituent in the North Sea. The amplitudes of such systems are a maximum along their coastal boundaries, and the phases rotate (either clockwise or anti-clockwise) such that high water on one side of the basin is balanced by low water on the other side. While these surface displacements propagate around the system in a tidal period, the net ebb or flood excursions of individual particles seldom exceeds 20 km.

These co-oscillating systems can accumulate energy over a number of cycles (see Section 2.5.4), resulting in spring tides occurring several days after new or full Moon. Basin morphology can selectively amplify the amphidromes for different constituents. In general, the observed amplitudes of semi-diurnal constituents relative to diurnal are significantly larger than indicated from their equilibrium ratios shown above.

### 1A.9  Monthly, fortnightly and quarter-diurnal constituents

In shallow water and close to abrupt changes in bathymetry, tidal constituents interact (see Section 2.6). From the trignometric relationship

$$\cos \omega_1 \cdot \cos \omega_2 = 0.5 \left( \cos(\omega_1 + \omega_2) + \cos(\omega_1 - \omega_2) \right), \qquad (1A.5)$$

a product of two constituents $\omega_1$ and $\omega_2$ results in constituents at their sum and difference frequencies. Thus, terms involving products of $M_2$ and $S_2$ generate constituents at the quarter-diurnal frequency $MS_4$ and the fortnightly frequency $MS_f$. Similarly, $M_2$ and $N_2$ generate constituents at the quarter-diurnal frequency $MN_4$ and the monthly Mm.

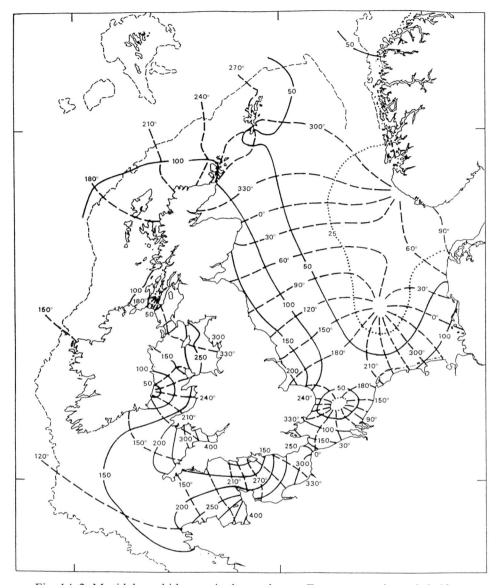

Fig. 1A.2. M$_2$ tidal amphidromes in the north west European continental shelf.

## References

Cartwright, D.E., 1999. *Tides: A Scientific History.* Cambridge University Press, Cambridge.
Dean, R.G., 1966. Tides and harmonic analysis. In: Ippen, A.T. (ed.), *Estuary and Coastline Hydrodynamics.* McGraw-Hill, New York, pp. 197–230.

Dyer, K.R., 1997. *Estuaries: A Physical Introduction*, 2nd ed. John Wiley, Hoboken, NJ.

Flather, R.A., 1976. A tidal model of the north west European Continental Shelf, *Memoires Societe Royale des Sciences de Liege, Ser,* **6** (10), 141–164.

Lane, A. and Prandle, D., 2006. Random-walk particle modelling for estimating bathymetric evolution of an estuary. *Estuarine, Coastal and Shelf Science,* **68** (1–2), 175–187.

Liu, W.C., Chen, W.B., Kuo, J-T, and Wu, C., 2008. Numerical determination of residence time and age in a partially mixed estuary using a three-dimensional hydrodynamic model. *Continental Shelf Research,* **28** (8), 1068–1088.

Prandle, D., 1982. The vertical structure of tidal currents and other oscillatory flows. *Continental Shelf Research,* **1**, 191–207.

Prandle, D., 2004. How tides and river flows determine estuarine bathymetries. *Progress in Oceanography,* **61**, 1–26.

Prandle, D., Lane, A., and Manning, A.J., 2005. Estuaries are not so unique. *Geophysical Research Letters,* **32** (23).

# 2

# Tidal dynamics

## 2.1 Introduction

Tidal propagation in estuaries can be accurately simulated using either numerical or hydraulic scale models. However, such models do not directly provide understanding of the basic mechanisms or insight into the sensitivities of the controlling parameters. Thus, while terms representing friction and bathymetry appear explicitly in (2.8) and (2.11), it is not immediately evident why tides are greatly amplified in certain estuaries yet quickly dissipated in others. The aim here is to derive analytical solutions, and thereby Theoretical Frameworks, to guide specific modelling and monitoring studies and provide insight into and perspective on estuarine responses generally.

Much of the theory developed here assumes that tidal propagation in estuaries can be represented by the shallow-water wave equations reduced to a 1D cross-sectionally averaged form. Section 2.2 describes the bases of this simplification. By further reducing these equations to a linear form, localised solutions are readily obtained, these are examined in Section 2.3.

It is shown in Section 2.4 that by introducing geometric expressions to approximate estuarine bathymetry, whole-estuary responses can be determined. Tidal responses in estuaries are shown for geometries approximated by (i) breadth and depth variations of the form $B_L(X/\lambda)^n$ and $H_L(X/\lambda)^m$, where $X$ is the distance from the head of the estuary, i.e. the location of the upstream boundary condition at the limit of tidal influence; (ii) breadth and depth varying exponentially and (iii) a 'synchronous' estuary. Chapters 6 and 7 provide details of 'synchronous estuaries', their geometry is shown to correspond to (i) with $m = n = 0.8$. By expressing the relevant equations in dimensionless form, these analytical solutions are transposed into Theoretical Frameworks, describing tidal elevations and currents over a wide range of estuarine conditions. Further details of current responses are described in Chapter 3.

Where a single ($M_2$) constituent predominates, this provides a robust basis for linearisation of the friction term as outlined in Section 2.5. Tidal propagation in estuaries often involves large excursions over rapidly varying shallow topography. While first-order tidal propagation is relatively insensitive to small topographic changes (Ianniello, 1979), Section 2.6 illustrates how the associated non-linearities result in the generation of significant higher harmonic and residual components with pronounced spatial gradients.

Finally, Section 2.7 indicates some of the peculiarities of surge–tide interactions.

## 2.2 Equations of motion

The equations of motion at any height $Z$ (measured vertically upwards above the bed) along orthogonal horizontal axes, $X$ and $Y$, may be written in Cartesian co-ordinates (neglecting vertical accelerations) as follows:

Accelerations in $X$-direction:

$$\frac{\partial U}{\partial t} + U\frac{\partial U}{\partial X} + V\frac{\partial U}{\partial Y} + g\frac{\partial \varsigma}{\partial X} - \Omega V = \frac{\partial}{\partial Z}E\frac{\partial U}{\partial Z} \qquad (2.1)$$

Accelerations in $Y$-direction:

$$\frac{\partial V}{\partial t} + U\frac{\partial V}{\partial X} + V\frac{\partial V}{\partial Y} + g\frac{\partial \varsigma}{\partial Y} + \Omega U = \frac{\partial}{\partial Z}E\frac{\partial V}{\partial Z} \qquad (2.2)$$

Continuity:

$$\frac{\partial U}{\partial X} + \frac{\partial V}{\partial Y} + \frac{\partial W}{\partial Z} = 0, \qquad (2.3)$$

where $U$, $V$ and $W$ are velocities along $X$, $Y$ and $Z$, $\varsigma$ is surface elevation, $\Omega = 2\omega \sin \varphi$ is the Coriolis parameter representing the influence of the earth's rotation ($\omega = 2\pi/24$ h), $\varphi$ is latitude and $E$ is a vertical eddy viscosity coefficient. Forcing due to wind and variations in density or atmospheric pressure is omitted in (2.1) and (2.2).

For many applications, it is convenient to vertically integrate between the bed and the surface. The depth-averaged equations retain the same form except that

(1) the non-linear convective terms $U\,(\partial U/\partial X) + V\,(\partial U/\partial Y)$ in (2.1) and $U\,(\partial V/\partial X) + V\,(\partial V/\partial Y)$ in (2.2) are multiplied by coefficients dependent on the vertical structure of $U$ and $V$; these coefficients are often assumed to equal 1 for simplicity;
(2) with zero surface stress, the vertical viscosity terms are replaced by bed stress terms $\tau_x/\rho D$ and $\tau_y/\rho D$, assumed to be proportional to the respective components of bed velocity squared, i.e.

$$\tau_x = -\rho f U(U^2 + V^2)^{1/2}, \tau_y = -\rho f V(U^2 + V^2)^{1/2}, \qquad (2.4)$$

where $\rho$ is water density and $f$ is the bed stress coefficient ($\approx 0.0025$)

(3) the kinematic boundary condition at the surface and bed is:

$$W_s = \frac{\partial \varsigma}{\partial t} + U \frac{\partial \varsigma}{\partial X} + V \frac{\partial \varsigma}{\partial Y}$$

and

$$W_0 = -U \frac{\partial D}{\partial X} - V \frac{\partial D}{\partial Y}. \tag{2.5}$$

These yield a depth-integrated continuity equation:

$$(D + \varsigma) \frac{\partial \varsigma}{\partial t} + \frac{\partial}{\partial X} U(D + \varsigma) + \frac{\partial}{\partial Y} V(D + \varsigma) = 0 \tag{2.6}$$

Ianniello (1977) indicates that transverse velocities can be neglected if the Kelvin Number $\Omega B/(gD)^{1/2} \ll 1$ and the horizontal aspect ratio $B^2\omega^2/(gD) \ll 1$ ($B$ breadth, $\omega = 2\pi/P$, $P$ tidal period). Thence adopting $X$ axially, by integrating across both breadth and depth, (2.1) may be rewritten in cross-sectionally averaged parameters as

$$\frac{\partial U}{\partial t} + U \frac{\partial U}{\partial X} + g \frac{\partial \varsigma}{\partial X} + f \frac{U|U|}{(D + \varsigma)} = 0 \tag{2.7}$$

and the continuity equation (2.6) as

$$B(D + \varsigma) \frac{\partial \varsigma}{\partial t} + \frac{\partial}{\partial X} BUA = 0, \tag{2.8}$$

where $A$ is the cross-sectional area.

Although lateral velocities may be restricted in estuaries, the transverse Coriolis term $\Omega U$, in (2.2), must be balanced, generally by a lateral surface gradient. This gradient produces an elevation phase advance on the right-hand side (looking landwards in the northern hemisphere) of the order of $B\Omega/(2(gD)^{1/2})$ radians (Larouche *et al.*, 1987).

The relative magnitudes of the terms in (2.7) for a predominant tidal frequency $\omega$ are approximately

$$\omega U^* : \frac{2\pi U^{*2}}{\lambda} : \frac{2\pi \varsigma^* g}{\lambda} : \frac{fU^{*2}}{D}, \tag{2.9}$$

where $\lambda$ is the wavelength over which both $U$ and $\varsigma$ vary. Assuming $\lambda = (gD)^{1/2}P$, the relative magnitudes of the first two terms are

$$(gD)^{1/2} : U^*. \tag{2.10}$$

Thus, the ratio of the magnitudes of the convective term and the temporal acceleration term is equal to the Froude number for the flow. This is generally small, and hence for first-order tidal simulation, the convective terms can be neglected. The relative magnitude of the friction term to the temporal acceleration term for the semi-diurnal frequency and $f = 0.0025$ is approximately $20\ U*/D\ \mathrm{s}^{-1}$, i.e. predominant in fast-flowing shallow estuaries (see Section 2.3.2). In such estuaries, the frictional force greatly exceeds the acceleration (inertial) term over most of the tidal cycle and wave propagation is diffusive in character (LeBlond, 1978).

## 2.3 Tidal response – localised

It is shown in Section 2.4.2 that combining the two equations, (2.7) and (2.8), produces an expression for tidal response along an estuary similar to the spectral response for a linearly damped, single-degree-of-freedom oscillatory system executing 'simple harmonic motion'. Thus, we expect harmonic solutions with axial variations in tidal amplitudes and phases described by Bessel functions, as illustrated in Section 2.4.1.

### 2.3.1 Linearised solution

Neglecting the convective term and linearising the friction term in (2.7) (see Section 2.5 for details of this linearisation) yields

$$\frac{\partial U}{\partial t} + g\frac{\partial \varsigma}{\partial X} + FU = 0. \tag{2.11}$$

It is readily shown that in a prismatic channel of infinite length and zero friction, (2.8) and (2.11) indicate a wave celerity, $c = (gD)^{1/2}$ and $U* = \varsigma(g/D)^{1/2}$ (Lamb, 1932). Maximum amplification then occurs for quarter-wave resonance at length $L = 0.25\lambda = 0.25\ P\ (gD)^{1/2}$. It is shown in Section 2.4.1 that even in damped, funnel-shaped estuaries, maximum amplification often occurs for values of $L$ close to this value.

Introducing a surface gradient for a predominant constituent in the form

$$\frac{\partial \varsigma}{\partial X} = \varsigma_x^* \cos \omega t \tag{2.12}$$

from (2.11), we obtain

$$U^* = \frac{-g\varsigma_x^*}{(F^2 + \omega^2)}(F\cos \omega t + \omega \sin \omega t). \tag{2.13}$$

Thus, for a frictionally dominated system $F \gg \omega$,

$$U^* = -g\varsigma_x^*/F\cos \omega t, \tag{2.14}$$

while for a frictionless system $F \ll \omega$,

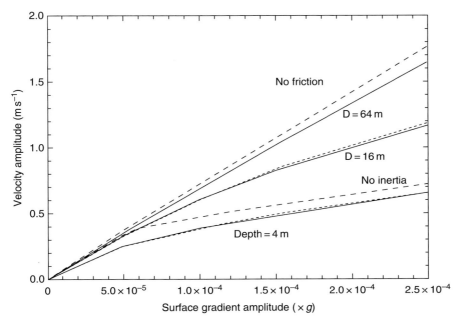

Fig. 2.1. $M_2$ tidal current amplitude as a function of surface gradient. Continuous lines show solution (2.13) for $D=4$, 16 and 64 m. Dashed lines show solution (2.14) for $D=4$ m and (2.15) for $D=64$ m.

$$U^* = -g\varsigma_x^*/\omega \sin \omega t. \tag{2.15}$$

Figure 2.1 indicates the solution of (2.13) for prescribed values of surface gradient up to 0.00025 g and depths of 4, 16 and 64 m. For the deepest case, the solution approximates (2.15) while the shallowest case approximates (2.14).

Figure 2.2 shows the magnitude of the terms in (2.7) at two positions in the Thames for the predominant $M_2$ constituent and the related higher harmonics $M_4$ and $M_6$, as calculated in a numerical model simulation (Prandle, 1980). For $M_2$, the inertial and frictional terms are orthogonal in phase and balance the surface gradient term. By contrast, for $M_4$ and $M_6$, the spatial gradient term is a consequence of rather than a driving force for currents (see Section 2.6) and hence different relationships apply.

### 2.3.2 Synchronous estuary solution

A 'synchronous estuary' is one where surface gradients associated with axial amplitude variations in $\varsigma^*$ are significantly less than those associated with corresponding phase variations. In deriving solutions to (2.8) and (2.11), a similar

*Tidal dynamics*

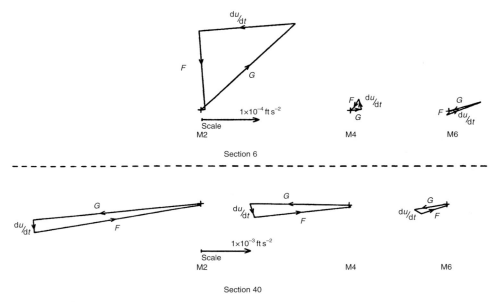

Fig. 2.2. $M_2$, $M_4$ and $M_6$ constituents of (2.7) near the mouth (top) and upstream in the Thames, based on numerical model simulation.

approximation is assumed to apply to axial variations in $U^*$. For the derived 'synchronous' bathymetry, these assumptions for both $U^*$ and $\varsigma^*$ have been shown to be valid, except in the shallowest conditions at the tidal limit (Prandle, 2003). Introduction of the solution here permits a ready comparison with other solutions and provides convenient expressions for $U^*$ in terms of $\varsigma^*$ and $D$ – subsequently used throughout this book.

Concentrating on the propagation of one predominant tidal constituent, $M_2$, the solutions for $U$ and $\varsigma$ at any location can be expressed as

$$\varsigma = \varsigma^* \cos(K_1 X - \omega t) \text{ and } U = U^* \cos(K_2 X - \omega t + \theta), \qquad (2.16)$$

where $K_1$ and $K_2$ are the wave numbers, $\omega$ is the tidal frequency and $\theta$ is the phase lag of $U$ relative to $\varsigma$.

Further assuming a triangular cross-section with constant side slopes, (2.8) reduces to

$$\frac{\partial \varsigma}{\partial t} + U\left(\frac{\partial \varsigma}{\partial X} + \frac{\partial D}{\partial X}\right) + \frac{1}{2}\frac{\partial U}{\partial X}(\varsigma + D) = 0. \qquad (2.17)$$

Friedrichs and Aubrey (1994) indicate that $U(\partial A/\partial X) \gg A(\partial U/\partial X)$ in convergent channels. Likewise, assuming $\partial D/\partial X \gg \partial \varsigma^*/\partial X$, we adopt the following form of the continuity equation:

$$\frac{\partial \varsigma}{\partial t} + U \frac{\partial D}{\partial X} + \frac{D}{2} \frac{\partial U}{\partial X} = 0. \qquad (2.18)$$

Substituting solution (2.16) into Eqs (2.11) and (2.18), four equations (pertaining at any specific location along an estuary) representing components of cos $\omega t$ and sin $\omega t$ are obtained. By specifying the synchronous estuary condition that the spatial gradient in tidal elevation amplitude is zero, the condition $K_1 = K_2 = k$ is derived, i.e. identical wave numbers for axial propagation of $\varsigma$ and $U$. Then, the following solutions for the amplitude, $U^*$, and phase, $\theta$, of tidal current together with bed slope, $SL = \partial D/\partial X$, are obtained:

$$\tan \theta = -\frac{F}{\omega} = \frac{SL}{0.5Dk}, \quad U^* = \varsigma^* g \frac{k}{(\omega^2 + F^2)^{1/2}}, \quad k = \frac{\omega}{(Dg/2)^{1/2}}. \qquad (2.19)$$

### Results

The above solutions are consistent with (2.13), the celerity $0.5 \, (gD)^{1/2}$ follows from the assumption of a triangular cross section. Chapter 6 illustrates how these explicit solutions for $U^*$, $\theta$ and $SL$ enable other related parameters, such as estuarine length, to be determined, yielding a range of Theoretical Frameworks in terms of the parameters $D$ and $\varsigma^*$. The parameter ranges selected are $\varsigma^*$ (0–4 m) and $D$ (0–40 m), representing all but the deepest of estuaries.

### Current amplitudes U

Figure 2.3 shows the solution (2.19) with current amplitudes extending to $1.5 \, \mathrm{m \, s^{-1}}$ (Prandle, 2004). The contours show that maximum values of $U^*$ occur at approximately $D = 5 + 10 \, \varsigma^*$ (m); however, these are not pronounced maxima. This figure explains why observed values of $U^*$ are so often in the range $0.5$–$1.0 \, \mathrm{m \, s^{-1}}$ despite large variations in $\varsigma^*$ over the spring–neap cycle and the wide range of estuarine depths.

### Role of bed friction

Friedrichs and Aubrey (1994) showed the predominance of the friction term in strongly convergent channels, irrespective of depth. Figure 2.4 shows the ratio of friction: inertia, $F/\omega$, from (2.19) (Prandle, 2004). $F/\omega$ is approximately equal to unity for $\varsigma^* = D/10$. For $\varsigma^* \ll D/10$, currents are insensitive to friction, while for $\varsigma^* \gg D/10$, tidal dynamics become frictionally dominated and currents decrease by a factor of two as the friction coefficient increases over its typical range from 0.001 to 0.004. Prandle (2003) provides a detailed analysis of the sensitivities to the friction parameter. From (2.19), for $F \gg \omega$, $U^* \propto \varsigma^{*1/2} D^{1/4} f^{-1/2}$ while for $F \ll \omega$, $U^* \propto \varsigma^* D^{-1/2}$.

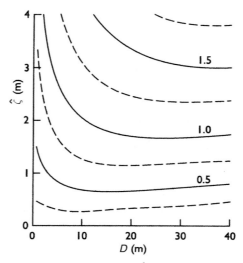

Fig. 2.3. Tidal Current amplitude, $U^*$ (m s$^{-1}$), as a $f(D, \varsigma^*)$ (2.19).

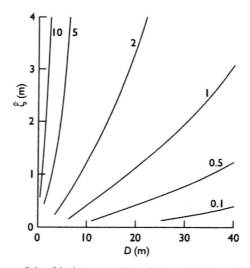

Fig. 2.4. Ratio $F : \omega$ of the friction term $F$ to the inertial term as a $f(D, \varsigma^*)$ (2.19).

From (2.19), $F/\omega = 0.1$ corresponds to a phase difference between tidal elevation and currents of $\theta = -6°$. Similarly, $F/\omega = 0.5$ corresponds to $\theta = -27°$, 1.0 to $-45°$, 2 to $-63°$, 5 to $-77°$ and 10 to $-84°$. These values of $\theta$ emphasise how the tidal wave propagation changes from 'progressive' in deeper water closer to the mouth to 'standing' in shallower water at the head.

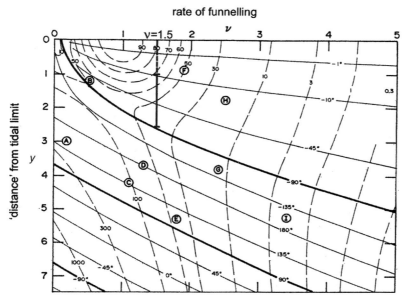

Fig. 2.5. Tidal elevation responses (2.24) for $s - 2\pi$. $\nu$ represents the degree of bathymetric funnelling and $y$ distance from the mouth, $y = 0$. Dashed contours indicate relative amplitudes, continuous contours relative phases. Vertical line at $\nu = 1.5$ shows typical lengths of synchronous estuaries (see Chapter 6). Lengths, $y$ (for $M_2$) and shapes, $\nu$, for estuaries (A)–(I) shown in Table 2.1.

### *Rate of funnelling in a synchronous estuary*

By integration of the solution for bed slope, $SL$, in (2.19), it can be shown that the synchronous solution corresponds to depths and breadths proportional to $X^{0.8}$, i.e. $m = n = 0.8$ in (2.20) and (2.21). Comparing the localised synchronous solutions to the whole-estuary response in Section 2.4.1, this synchronous geometry corresponds to $\nu = 1.5$. From Fig. 2.5, this is close to the centre of the range of geometries encountered (Prandle, 2004). Moreover, the estuarine lengths determined for synchronous estuaries, incorporated in Fig. 2.5, range from a small fraction up to close to that for 'quarter-wavelength' (first node), resonance at the $M_2$ frequency.

### **2.4 Tidal response – whole estuary**

This section is concerned with the first-order response of whole estuaries to tidal forcing. The aim is to construct simplified analytical frameworks that answer such basic questions as to why (i) tides are large in some estuaries, (ii) semi-diurnal constituents are sometimes amplified while diurnals are often damped and (iii) some estuaries are sensitive to small changes in bed friction, length or depth. Simplified analytical solutions to (2.8) and (2.11) have been presented by Taylor (1921), Dorrestein (1961) and Hunt (1964). Here we discuss more generalised solutions for (i) breadth and depth varying with powers of distance $X$ (Prandle and Rahman, 1980, subsequently Prandle and Rahman, 1980) and (ii) breadth and depth varying exponentially with $X$ (Prandle, 1985). Since these responses are based on the linearised equations, they are generally applicable to estuaries with a predominant tidal constituent.

Taylor's frictionless solution for an estuary with linearly varying depth and breadth represents a special case of (i). Hunt's analytical solutions for estuaries with exponentially increasing breadth and constant depth are presented in Section 2.4.2.

### 2.4.1  Breadth and depth varying with powers of distance X (Prandle and Rahman, 1980)

Breadth and depth are assumed to vary by

$$B(X) = B_{\text{L}}\left(\frac{X}{\lambda}\right)^n \tag{2.20}$$

and

$$H(X) = H_{\text{L}}\left(\frac{X}{\lambda}\right)^m \tag{2.21}$$

with $X$ measured from the head of the estuary. To convert to a dimensionless format, we adopt $\lambda$ as a unit of horizontal dimension, $H_{\text{L}}$ as a unit of vertical dimension and $P$, the tidal period, as a unit of time, with

$$\lambda = (gH_{\text{L}})^{1/2}P \tag{2.22}$$

corresponding to the tidal wavelength for $H_{\text{L}}$ constant. Dimensionless parameters are introduced as follows:

$$x = X/\lambda, t = T/P, h = H/H_{\text{L}}, b = B/\lambda, u = UP/\lambda, \text{ frictional parameter } s = FP. \tag{2.23}$$

Prandle and Rahman (1980) showed that the substitution of (2.20) and (2.21) into (2.8) and (2.11) yields the following solution for tidal elevation $\varsigma$ at any location $x$, at any time $t$, for any tidal period $P$:

$$\varsigma = \varsigma^* \left(\frac{ky}{ky_{\text{M}}}\right)^{1-v} \frac{\mathbf{J}_{v-1}(ky)}{\mathbf{J}_{v-1}(ky_{\text{M}})} e^{\mathrm{i}2\pi t}, \tag{2.24}$$

where $\varsigma^* e^{\mathrm{i}2\pi t}$ is the tidal elevation at the mouth $x_{\text{M}}$ and

$$v = \frac{n+1}{2-m}, \quad k = \left(\frac{1-\mathrm{i}s}{2\pi}\right)^{1/2}$$

$$y = \frac{4\pi}{2-m} x^{\frac{2-m}{2}} \tag{2.25}$$

and $J_{v-1}$, is a Bessel function of the first kind and of order $v-1$.

The solution (2.24) is illustrated in diagrammatic form in Fig. 2.5 for the case of $s = 2\pi$, i.e. $F = \omega$. Prandle and Rahman (1980) show the corresponding solutions for $s = 0.2\pi$. Away from the resonant conditions illustrated in Fig. 2.6, the responses for the two frictional coefficients are essentially similar, with reduced amplitudes and enhanced phase differences for the larger friction coefficient. This figure constitutes a general response diagram showing the variation in amplitude and phase of tidal elevations along the length of an estuary. For the $M_2$ semi-diurnal constituent, the positions indicated (A)–(I) designate the mouths of the major estuaries listed in Table 2.1.

Confidence in the validity of this approach was shown by comparing results for $M_2$ elevation response for the ten major estuaries listed in Table 2.1. In all of these

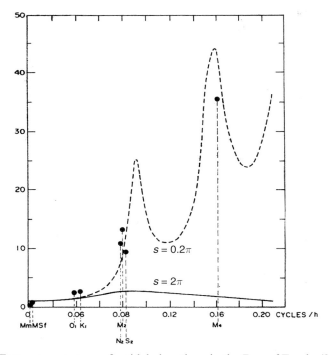

Fig. 2.6. Frequency response for tidal elevations in the Bay of Fundy (2.24), with $s = 0.2\pi$ and $2\pi$. Vertical scale shows amplification at the head relative to values at the shelf edge encircled dots represent observed data.

Table 2.1 *Geometrical parameters for ten estuaries shown in Fig. 2.5*

|   | $H_M$ (m) | $L$ (km) | $n$ | $m$ | $v$ | $y_0$ | $H_0$ (m) | $\alpha$ | $\beta$ | $\alpha + 2\beta$ |
|---|---|---|---|---|---|---|---|---|---|---|
| A Fraser | 44 | 135 | −0.7 | 0.7 | 0.2 | 3.0 | 2.3 | −2.8 | 2.8 | 2.8 |
| B Rotterdam Waterway | 13 | 99 | 0 | 0 | 0.5 | 1.2 | 13.0 | 0 | 0 | 0 |
| C Hudson | 17 | 248 | 0.7 | 0.4 | 1.1 | 4.2 | 4.8 | 2.2 | 1.3 | 4.8 |
| D Potomac | 13 | 184 | 1.0 | 0.4 | 1.3 | 3.7 | 3.5 | 3.6 | 1.4 | 6.4 |
| E Delaware | 5 | 214 | 2.1 | 0.3 | 1.8 | 5.3 | 2.3 | 5.3 | 0.8 | 6.9 |
| F Miramichi | 7.0 | 55 | 2.7 | 0 | 1.9 | 0.9 | 7.0 | 46.6 | 0 | 46.6 |
| G Bay Fundy | 2000 | 635 | 1.5 | 1.0 | 2.4 | 3.8 | 21.4 | 3.9 | 2.6 | 9.1 |
| H Thames | 80 | 95 | 2.3 | 0.7 | 2.5 | 1.77 | 2.7 | 14.1 | 4.3 | 22.7 |
| I Bristol Channel | 5000 | 623 | 1.7 | 1.2 | 3.4 | 5.20 | 12.5 | 3.4 | 2.4 | 8.2 |
| J St. Lawrence | 300 | 418 | 1.5 | 1.9 | 19.5 |  | 1 | 1.3 | 1.6 | 4.5 |

*Notes:* $H_M$ depth at mouth, $H_0$ depth at head, $L$ and $y_0$ estuarine lengths (from (2.25))
$n$, $m$, $v$, $\alpha$ and $\beta$ bathymetric parameters.
*Source:* Prandle and Rahman, 1980; Prandle, 1985

estuaries, good agreement was found using $s = 2\pi$, except for the Bay of Fundy (G), where with depths exceeding 200 m better agreement was found for $s = 0.2\pi$.

Given the funnelling factor $v$ and length $y_M$ for a particular estuary, the variation in amplitude and phase in the estuary can be read along the corresponding vertical line. Moreover, the value of $y_M$ is inversely proportional to the tidal period $P$, thus doubling $P$ halves $y_M$. Using this relationship, Fig. 2.6 illustrates the spectral response for the Bay of Fundy. Strictly, some adjustment is necessary to reflect the 50% increase in the friction factor appropriate to constituents other than the predominant $M_2$ – as described in Section 2.5. Similar response diagrams to Fig. 2.5 can be constructed for amplitudes and phases of tidal currents.

In summary, Fig. 2.5 constitutes a general tidal response diagram indicating amplitudes and phases (relative to the mouth) at all positions, for all tidal periods, for all estuaries which reasonably correspond to (2.11) and (2.12). This response diagram explains a number of features commonly encountered:

(1) The quarter-wavelength resonance or primary mode found in sufficiently long estuaries is indicated by the thick line through the amplitude nodes.
(2) For a diurnal tidal constituent, the $y_M$ values (A)–(I) are halved; hence, we expect relatively small amplification of such constituents. For $MS_f$, a 14-day constituent, the reduction in the $y_M$ values would indicate little amplification or phase difference along any estuary.
(3) For quarter-diurnals or other higher harmonics, in relative terms we expect high amplification, large phase differences and one or more nodal positions. However, it is important to distinguish between the response to external forcing represented by the present analysis versus the internal generation of higher harmonics by non-linear processes within an estuary discussed in Section 2.6 and illustrated in Fig. 2.2.

Table 2.2 *Resonant lengths for funnel-shaped estuaries,* $B \propto X^n$ *and* $D \propto X^m$, *(2.26), as a fraction of the prismatic value (2.22)*

| $m =$ | | | |
|---|---|---|---|
| $n$ | 0 | 0.8 | 1.0 |
| 0 | 1.04 | 0.74 | 0.57 |
| 0.8 | 1.23 | 0.91 | 0.83 |
| 2.5 | 1.64 | 1.33 | 1.26 |

*Quarter-wavelength resonance*

The position of the first nodal line in Fig. 2.5, corresponding to maximum amplification in estuaries with varying degrees of funneling, is approximated by

$$y = 1.25 + 0.75v, \quad \text{i.e. } x = \left(\frac{3n - 5m + 13}{16\pi}\right)^{(2/2-m)}. \tag{2.26}$$

The ratio of resonant lengths, (2.26), as a fraction of that for a prismatic channel, (2.22), is $x^{(1-m/2)}/0.25$.

Table 2.2 shows this ratio, for the typical ranges of $0 < n < 2.5$ and $0 < m < 1$, including the value $m = n = 0.8$ corresponding to the solution for a synchronous estuary (Section 2.3.2). These results indicate that upstream reductions in depths decrease resonant lengths while breadth convergence has the opposite effect, emphasising the complexity of tidal response in funnel-shaped estuaries.

### 2.4.2 Depth and breadth varying exponentially (Hunt, 1964; Prandle, 1985)

Assuming a breadth and depth variation:

$$B(X) = B_0 \exp(nX)$$
$$H(X) = H_0 \exp(mX), \tag{2.27}$$

where $B_0$ and $H_0$ are the respective values at the head $X = 0$, we convert to dimensionless units based on the tidal period $P$ as the unit of time. $H_0$ as the vertical dimension and $\lambda$ the horizontal dimension given by

$$\lambda = (gH_0)^{1/2}P. \tag{2.28}$$

Thus, we obtain transformed dimensionless variables: $x = X/\lambda$, $t = T/P$, $z = Z/H_0$, $b = B/\lambda$, $h = H/H_0$, $u = U(P/\lambda)$, $s = FP$ and

$$b(x) = b_0 \exp(\alpha x) \tag{2.29}$$

and

$$h(x) = \exp(\beta x), \tag{2.30}$$

where $\alpha = n\lambda$ and $\beta = m\lambda$.

Substituting (2.29) and (2.30) into (2.8) and (2.11), these equations may be rearranged to form separate expressions for either $\varsigma$ or $u$. The time derivatives in these expressions may be eliminated by considering amplitudes pertaining to a single period $P$, thus

$$\varsigma = \varsigma^* \exp(\mathrm{i}2\pi t) \text{ and } u = u^* \exp(\mathrm{i}2\pi t). \tag{2.31}$$

The expressions for the tidal amplitudes $\varsigma$ and $u$ are then as follows:

$$\frac{\partial^2}{\partial x^2}\varsigma^* + (\alpha + \beta)\frac{\partial \varsigma^*}{\partial x} + (4\pi^2 - 2\pi\mathrm{i}s)\frac{\varsigma^*}{\exp(\beta x)} = 0 \tag{2.32}$$

$$\frac{\partial^2 u^*}{\partial x^2} + (\alpha + 2\beta)\frac{\partial u^*}{\partial x} + \left[\beta(\alpha + \beta) + \frac{(4\pi^2 - 2\pi\mathrm{i}s)}{\exp(\beta x)}\right]u^* = 0. \tag{2.33}$$

By introducing appropriate transformations, the middle terms (involving the single derivative in $x$ in (2.32) and (2.33)) may be eliminated. The resulting equations may then be solved analytically (Gill, 1982, Section 8.12). Such solutions have been examined by Xiu (1983); however, their complexity obscures direct understanding. In the following section, we consider simpler analytical solutions relating to certain special cases alongside a numerical solution to illustrate the nature of the responses.

**(1)** Solutions for constant depth, $\beta = 0$: Hunt (1964) showed that for this case the solutions to (2.32) and (2.33) are

$$\varsigma = \varsigma^*{}_0 \exp\left(\frac{-\alpha\,x}{2}\right)\left(\cosh \omega x + \frac{\alpha}{2\omega}\sinh \omega x\right) \tag{2.34}$$

$$u = -\varsigma^*{}_0 \exp\left(\frac{-\alpha\,x}{2}\right)\frac{2\pi\mathrm{i}}{\alpha}\sinh \omega x, \tag{2.35}$$

where $\omega = \omega_1 + \mathrm{i}\,\omega_2$, $\omega_1{}^2 - \omega_2{}^2 = \alpha^2/4 - 4\pi^2$, $\omega_1 \cdot \omega_2 = \pi s$ and $\varsigma^*{}_0$ is the elevation amplitude at the head, $x = 0$.

**(2)** Solutions for constant depth, $\beta = 0$ and zero friction: For this case, (2.32) and (2.33) described the free vibrations of a damped simple harmonic oscillator. Using this analogy, for $\alpha < 4\pi$ the system is under-damped, $\alpha = 4\pi$ represents critical damping while $\alpha > 4\pi$ is over-damped.

For $\alpha > 4\pi$, the solutions retain the form shown in (2.34) and (2.35) with $\omega_1^2 = \alpha^2/4 - 4\pi^2$ and $\omega_2 = 0$.

For $\alpha < 4\pi$, the solutions simplify to

$$\varsigma^* = \varsigma^*{}_0 \exp\left(\frac{-\alpha x}{2}\right)\left(\cos \omega_2 x + \frac{\alpha}{2\omega_2}\sin \omega_2 x\right) \tag{2.36}$$

$$u^* = -\varsigma^*_0 \exp\left(\frac{-\alpha x}{2}\right) \frac{2\pi i}{\omega_2} \sin \omega_2 x \qquad (2.37)$$

with $\omega_2{}^2 = -\alpha^2/4 + 4\pi^2$.

For $\alpha = 4\pi$, the following specific solutions apply

$$\varsigma^* = \varsigma^*_0 \exp(-2\pi x)(1 + 2\pi x) \qquad (2.38)$$

$$u^* = \varsigma^*_0 \exp(-2\pi x) 2\pi i x. \qquad (2.39)$$

**(3)** Numerical solution for both depth and breadth varying exponentially and friction.

The general response diagram is shown in Fig. 2.7 (Prandle, 1985), for $s = 2\pi$, the orthogonal axes refer to the parameters $\alpha$ and $\beta$. The contours show the amplification between the amplitude of the tidal elevation at the head of the estuary relative to the value at the first nodal position. However, for estuaries with values of $\alpha + 2\beta > 10$, no nodal position occurs and in this case the amplification shown is relative to the value for $x = 1$ where the latter value closely approximates the asymptote at $x = \infty$. This demarcation in the response of estuaries at $\alpha + 2\beta = 10$ was not evident in the solution in Section 2.4.1. The symbols (A)–(J) again indicate the amplification for all ten major estuaries, listed in Table 2.1, between the head and the first nodal position or $x = 1$ (not the mouth) for the semi-diurnal constituent $M_2$.

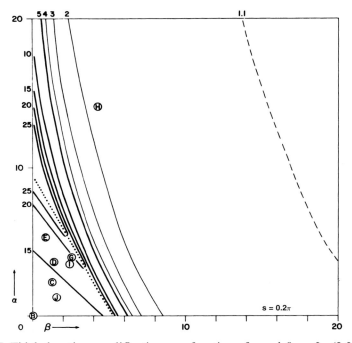

Fig. 2.7. Tidal elevation amplification as a function of $\alpha$ and $\beta$, $s = 2\pi$ (2.32).

   To determine the maximum response for other tidal constituents from Fig. 2.7, we note that the values for $\alpha$ and $\beta$ are directly proportional to period; thus, for diurnal constituents $\alpha$ and $\beta$ are doubled while for quarter-diurnal constituents $\alpha$ and $\beta$ are halved. In consequence, we may deduce the following conclusions from the general response diagram, Fig. 2.7 and the analytical solutions in Section 2.4.2.

(a) The response of any estuary may be likened to the free vibrations of a damped simple harmonic oscillator. Estuaries of type I with $\alpha + 2\beta < 10$ are under-damped and elevation amplitudes vary in an oscillatory manner along the $x$-axis. Estuaries of type II with $\alpha + 2\beta \sim 10$ are critically damped and produce maximum amplification. Estuaries of type III with $\alpha + 2\beta \gg 10$ are over-damped and elevations increase monotonically towards the head with little amplification and little sensitivity to frictional effects. The value of 10 is approximate, the corresponding value of $4\pi$ was determined for the zero-friction case in Section 2.4.2(2).

(b) The Bristol Channel (I) and Bay of Fundy (G) both lie close to the demarcation line, $\alpha + 2\beta \sim 10$, in Fig. 2.7, explaining the sensitive 'near-resonant' nature of the response of these two systems. These responses are a consequence of their particular values of $\alpha$ and $\beta$ (for the $M_2$ constituent) and not simply due to their resonant lengths as generally assumed.

## 2.5 Linearisation of the quadratic friction term

### 2.5.1 Single constituent

The preceding analyses required the quadratic friction term $f_1 U|U|$ to be approximated by a linear term $f_2 U$. For a single constituent $U = U_1 \cos \omega t$, equating energy dissipation, proportional to $U^3$, over a tidal cycle requires $f_2 = (8/3\pi) f_1 U_1$.

### 2.5.2 Two constituents

Proudman (1923) and Jeffreys (1970) showed that where two tidal stream constituents $U_1 \cos \omega t$ and $U_1' \cos \omega' t$ coincide, the friction term is given by

$$f_1(U_1 \cos \omega t + U_1' \cos \omega' t)|U_1 \cos \omega t + U_1' \cos \omega' t|. \tag{2.40}$$

Then, for small values of $U'_1$, the linearised friction component $F_\omega$ at the frequency of $\omega$ is given, to a first approximation, by

$$F_\omega = \left(\frac{8f_1}{3\pi}\right) U_1^2 \cos \omega t = f_2 U_1 \cos \omega t \tag{2.41}$$

while the component at the frequency $\omega'$ is

$$F_{\omega'} = \left(\frac{4}{\pi}\right) f_1 U_1 U_1' \cos \omega' t = \left(\frac{3}{2}\right) f_2 U_1' \cos \omega' t. \tag{2.42}$$

In the case of a small residual current $U_0$ coincident with a large tidal stream $U_1 \cos \omega t$, Bowden (1953) showed that, with $U_1 \gg U_0$, the linear friction term associated with the residual current is given by

$$F_0 = \left(\frac{4}{\pi}\right) f_1 \, U_1 \, U_0 \; = \; \left(\frac{3}{2}\right) f_2 \, U_0. \tag{2.43}$$

Hence, (2.41), (2,42) and (2.43) show that to account for the interaction with a predominant $M_2$ constituent in a linearised model, when simulating separately any other tidal constituent or a residual flow, the linearised frictional coefficients $f_R$ must be computed according to

$$f_R = \left(\frac{3}{2}\right) f_2. \tag{2.44}$$

The magnitude of the tidal streams due to $M_2$ in mid-latitudes is typically of order two or three times the magnitude of the next largest constituent, $S_2$. Hence, it is reasonable, in simulating $M_2$ alone, to neglect the frictional interaction due to other constituents and use (2.41). However, in simulating any other tidal constituent, it is appropriate to linearise the frictional term by reference to the tidal velocities associated with $M_2$ by using (2.44). Hence, relative to other terms in the momentum equation, the linearised friction coefficient for other constituents, $\omega'$, is a factor 1.5 $U_{M_2}/U_{\omega'}$ larger. In the north west shelf seas, $S_2$ amplitudes are typically 0.33 of $M_2$, whereas the ratio of their 'equilibrium' potentials is 0.46, (Appendix 1A). Thus, increasing the linearised $S_2$ friction factor by a factor of 4.5 (relative to that for propagation of $M_2$) appears to reduce $S_2$ amplitudes by about 1/3.

Garrett (1972), Hunter (1975) and Saunders (1977) discuss the extension of the above results for flow in two dimensions.

### 2.5.3 Triangular cross section

For the synchronous solution, Section 2.3.2, the component of $f(U|U|/D)$ at the predominant tidal frequency $M_2$ was approximated by

$$\frac{8}{3\pi} \frac{25}{16} f \frac{|U^*|U}{D} = FU, \tag{2.45}$$

i.e. $F = 1.33 \, f \, U^*/D$, where $8/3\pi$ derives from the linearisation of the quadratic velocity term described above. The factor 25/16 is included to weight the frictional term close to the largest flow which occurs in the deepest water where, from (2.19) with $F \gg \omega$, the maximum velocity is 5/4 of the cross-sectional mean.

### *2.5.4  Q factor (Garrett and Munk, 1971)*

Analogies between the propagation of tides in a channel and the transmission of electrical energy in an AC circuit were made by Van Veen (1947) and Prandle (1980). The analogy links tidal elevations to voltage, tidal streams to current, bed friction to resistance and, somewhat less directly, inertial effects to inductance and surface area to capacitance. The analogy is dependent on the assumption that tidal propagation is sensibly linear and lightly damped. Evidence that tidal propagation is essentially linear in oceans and shelf seas follows from the accuracy of tidal prediction techniques, i.e. the harmonic method and, in particular, the response method (Munk and Cartwright, 1966).

The relative influence of the friction term in (2.11) was discussed in Section 2.3. Using the above analogy with an AC electrical circuit, a succinct quantification of frictional influence is provided by the quantity $Q$, known as the quality factor ($Q$ factor). An oscillating system dissipates $2\pi/Q$ of its energy each cycle. For systems near resonance,

$$Q = \frac{\omega_0}{\omega_2 - \omega_1} \tag{2.46}$$

is a measure of the sharpness of resonance, based on the ratio of the natural frequency, $\omega_0$, to the frequency difference between points on the response curve, $\omega_2$ and $\omega_1$, corresponding to half the power dissipation of that at the natural frequency.

Godin (1988) showed that for (2.11), the $Q$ factor of tidal basins is given by $Q = \omega/F$. Thus, the value of $s = FP = 2\pi$ used in reproducing the responses of estuaries in Section 2.4 is equivalent to $Q = 1$ and emphasises that most estuaries are highly dissipative. Conversely, in Fundy, for which $s = 0.2\pi$, $Q = 10$, indicating a more sharply resonant system. These estuarine values for $Q$ may be compared with the value of $Q \approx 17$ computed for the North Atlantic by Garrett and Greenberg (1977).

In 'spinning-up' a numerical simulation of tidal propagation starting from still-water conditions, it similarly follows from (2.11) that cyclical convergence of tidal amplitudes is approached asymptotically at a rate $(1 - \exp(-Ft))$.

### 2.6  Higher harmonics and residuals

The above theories provide robust descriptions of the first-order estuarine responses for the primary tidal constituents. However, subsequent chapters emphasise the longer-term importance to both mixing processes and sediment dynamics of seemingly 'second-order' effects, namely higher-order and residual tides alongside vertical, lateral and high-frequency variability in currents and salinity. While the first-order effects can be accurately modelled, numerical simulation of these 'second-order' effects requires increasingly fine resolution in both space and time. Thus, ironically, despite the rapid growth in computer power since the first simple, but successful, numerical tidal models, limitations in computing power remain an obstacle to progress. Here, we explain how these higher harmonic and residual terms are generated.

### 2.6.1 Trigonometry

Although the tidal forcing at the mouth of an estuary is primarily at the semi-diurnal or diurnal frequencies, the non-linear terms in (2.1) to (2.5) almost always produce significant higher harmonics in shallow water (Aubrey and Speer, 1985). This process can be understood from simple trigonometry in which

$$U^*_1 \cos(\omega_1 t) \times U^*_2 \cos(\omega_2 t) = \frac{U^*_1 U^*_2}{2} (\cos(\omega_1 - \omega_2)t + \cos(\omega_1 + \omega_2)t).$$

$$(2.47)$$

Thus, terms involving a product of the predominant $M_2$ constituent generate both $M_4$ and $Z_0$ constituents. Similarly, whenever two large constituents are present (e.g. $M_2$ and $S_2$), the same mechanisms generate constituents at their sum frequency $(\omega_1 + \omega_2) = \omega_{MS_4}$ and the difference frequency $(\omega_2 - \omega_1) = \omega_{MS_f}$, i.e. quarter-diurnal and fortnightly periods, or for $(M_2 + N_2)$, quarter-diurnal and monthly periods. These residuals and higher harmonics manifest themselves via features such as asymmetry between ebb and flood flows (particularly $M_4$), occasional double high waters and tidal pumping (in which an estuary exchanges water over a neap–spring cycle complicating mass-balance calculations based on observations made over a single semi-diurnal period). In formulating models to examine these residual and higher harmonics, it should be recognised that the scaling arguments used to derive (2.11), based on the predominant constituent, may be invalidated. In particular, for higher harmonics, the wavelength $\lambda$ associated with the variability of $U$ and $\varsigma$ may be determined by bathymetry (Zimmerman, 1978).

### 2.6.2 Non-linear terms in the tidal equations

A convenient device to illustrate the nature of these non-linearities is to rewrite (2.1) and (2.6) in terms of mass transports. These equations then take the following form (Prandle, 1978):

$$\frac{\partial}{\partial t} Q_x + \frac{\partial}{\partial X} \frac{Q_x^2}{H} + \frac{\partial}{\partial X} \frac{Q_x Q_y}{H} + Hg \frac{\partial \varsigma}{\partial X} + g \frac{f Q_x (Q_x^2 + Q_y^2)^{1/2}}{H^2} - \Omega Q_y = 0$$

$$(2.48)$$

$$\frac{\partial}{\partial t}\varsigma + \frac{\partial}{\partial X}Q_x + \frac{\partial}{\partial Y}Q_y = 0, \tag{2.49}$$

where $Q_x = UH$, $Q_y = VH$ and $H = D + \varsigma$.

Since (2.49) is linear, it is sufficient to consider only (2.48). By restricting the analysis to the consideration of the propagation of a single tidal constituent, the following assumptions may be introduced:

$$U = U_0 + U_1 \cos(-\omega t + \theta), \quad V = V_0 + V_1 \cos(-\omega t + \psi) \text{ and}$$
$$\varsigma = \varsigma_0 + \varsigma_1 \cos(-\omega t), \tag{2.50}$$

where the parameters $U_0$, $V_0$ and $\varsigma_0$ denote residuals. In the following analysis, it is also assumed that

$$U_0 \ll U_1, V_0 \ll V_1 \text{ and } \varsigma_0 \ll \varsigma_1. \tag{2.51}$$

### Inertial term

Incorporating (2.50) and integrating the first term in (2.48) over a tidal cycle gives a residual flow:

$$Q_0 = U_0 D + 0.5 \, U_1 \varsigma_1 \cos \theta. \tag{2.52}$$

This shows that, except in the case of $\theta = -\pi/2$ for a standing wave, the residual flow comprises both a term associated with net residual current and a second term known as the Stokes' transport. In a closed estuary with negligible river flow, a current, $U_0$, must balance the Stokes' drift. In the case of a purely progressive wave, $\theta = 0$, $U_0 = -05 U_1 \varsigma_1 / D$, i.e. seawards.

### Convective term

By expanding the second, convective, term in (2.48), it may be shown that, to a first approximation:

$$\frac{\partial}{\partial X}\frac{Q_x^2}{H} \approx \frac{\partial}{\partial X}\left(\frac{DU_1^2}{2}\right)(\cos -2\omega t + 1). \tag{2.53}$$

Thus, this convective term associated with the predominant constituent frequency, $\omega$, generates both a residual (steady component) and a constituent with frequency $2\omega$, i.e. $M_2$ generates $Z_0$ and $M_4$.

Similarly, expanding the third, convective, term in (2.48),

$$\frac{\partial}{\partial Y}\frac{Q_x Q_y}{H} \approx \frac{\partial}{\partial Y}\left(\frac{DU_1 V_1}{2}\right)(\cos -2\omega t + 1) \tag{2.54}$$

with similar results to (2.53).

Since, from (2.13), tidal current amplitudes adjust rapidly to changes in bathymetry, both (2.53) and (2.54) emphasise how bathymetric changes can generate pronounced non-linearities in tidal propagation.

### *Surface gradient*

Expanding the fourth term in (2.48) produces two distinct, residual terms:

$$\overline{Hg\frac{\partial \varsigma}{\partial X}} \approx g(D + \varsigma_0)\frac{\partial \varsigma_0}{\partial X} + \frac{1}{2}g\varsigma_1\frac{\partial \varsigma_1}{\partial X}. \tag{2.55}$$

These terms are not obtained explicitly in the corresponding analysis of residual terms associated with (2.7). In the present case, where the equation is in transports, the terms may be considered to represent the non-linearities due to shallow-water effects in the continuity equation (2.3). The first of the two terms represents the variation in msl. Nihoul and Ronday (1975) described the second term as 'tidal radiation stress' – analogous to 'set-up' in wind waves. In an application in the southern North Sea, they found that the first and second terms were of the same order of magnitude.

### *Quadratic friction*

The modulus in the friction term prevents simple trigonometric expansion, but it can be shown (Cartwright, 1968) that

$$U^{*2} \sin \omega t |\sin \omega t| = \left(\frac{8}{3\pi}\right) U^{*2} (\sin \omega t - (1/5) \sin 3\omega t - (1/35) \sin 5\omega t \ldots.). \tag{2.56}$$

Thus, the quadratic friction term generates odd harmonics (i.e. $M_6$, $M_{10}$ etc from $M_2$). However, approximating $1/(\varsigma + D)$ in the quadratic bed stress term by $(1 - \varsigma/D)/D$, we see that the combination of the $\varsigma/D$ term with the $M_2$ current in (2.56) can also lead to significant contributions at the frequency $M_4$ from the friction term.

### *Coriolis*

While the Coriolis term is linear and does not generate residual flows, it can play a major role in configuring the residuals produced by the non-linear terms.

### 2.6.3 *Ebb–flood asymmetry*

Friedrichs and Aubrey (1988) have shown how ebb and flood asymmetries generate net residual velocities and associated net differential erosion potentials. Tidal rectification associated with the phase-locked $M_2$ and $M_4$ constituents is of particular interest.

Sediment erosion and deposition are related more directly to near-bed velocities than to the magnitude and direction of (depth-integrated) flows. Hence, the relevant parameter generating non-linearities is current rather than flow. In many estuaries, the cross-sectional area at low water is a small fraction of that at high water. Thus, propagation of an oscillatory flow constituent introduces major non-linearities in currents.

Assuming a triangular cross section with constant side slopes, $\tan \alpha$, continuity of a sinusoidal net ebb and flood flow requires

$$U_1(t)A = \frac{U_1(t)[\varsigma_1(t) + D]^2}{\tan \alpha} + [U_2(t) + U_0(t)]A, \qquad (2.57)$$

where the cross-sectional area at mean water level $A = D^2/\tan \alpha$. The currents $U_2$ and $U_0$ are the first higher-harmonic and residual current components required to balance the oscillatory flows associated with the predominant constituent given by

$$\varsigma_1(t) = \varsigma_1^* \cos(-\omega t) \quad \text{and} \quad U_1(t) = U_1^* \cos(-\omega t + \theta). \qquad (2.58)$$

Then retaining only terms of $O(a)$, where $a = \varsigma_1^*/D$, we obtain

$$U_2(t) = -U_1^* a \cos(-2\omega t + \theta) \text{ and } U_0 = -U_1^* a \cos(\theta). \qquad (2.59)$$

Equation (2.59) indicates that a net downstream current accompanies the propagation of the primary tidal constituent. In shallow, macro-tidal estuaries, these two terms are likely to be the most significant non-linear current components.

## 2.7 Surge–tide interactions

Detailed descriptions of the generation and propagation of storm surges are beyond the present scope, see Heaps (1967, 1983) for further information. However, the following example of surge–tide interaction is included to illustrate the potential magnitude and complexity of interactions when the component terms are of similar magnitude and 'period'.

Flooding often involves not only large but 'peculiar' surges. The surge threat to London arises when the peak of a large storm surges coincides with the peak of high water on a spring tide. However, maximum surge peaks, defined as the difference between observed and (tidally) predicted water levels, invariably occur on the rising tide, a few hours before forecasted high water. However, reliance on this statistical relationship, albeit robust, seemed too precarious and the Thames Flood Barrier was constructed to protect London.

Figure 2.8 shows results from a numerical solution of surge–tide interaction in the Thames for the flooding of 1970 (Prandle and Wolf, 1978). A conceptual division

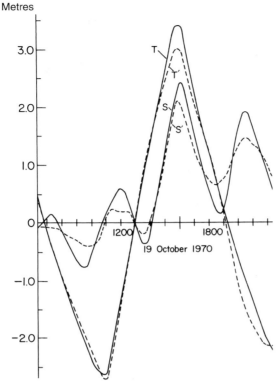

Fig. 2.8. Surge-tide interaction at Tower Pier in the Thames. *S* surge, *T* tide, *S'* and *T'* components with interaction terms cross linked.

into separate components of interaction was used to link simultaneous 'parallel' simulations of tide and surge. Dynamical coupling was introduced via the non-linear terms. Thus, for example $U_S + U_T \mid U_S + U_T \mid$ in the quadratic friction term was represented by the component $U_S \mid U_S + U_T \mid$ in the surge model (subscript S) and $U_T \mid U_S + U_T \mid$ in the tidal model (subscript T).

Using this 'parallel' model approach, Prandle and Wolf (1978) showed how this systematic occurrence of surge peaks on the rising tide is due to the advance of the tidal phase as a result of the displacement of the $M_2$ amphidromic system in the adjacent southern North Sea (Fig. 1A.2). This displacement is caused by enhanced depths due to positive surge levels throughout the North Sea. The subsequent reduction in surge amplitude immediately before the highest water level shown in Fig. 2.8 is mainly due to enhanced frictional dissipation. This enhancement is concentrated in space and time within a shallow coastal region in the outer Thames approaches and occurs when maximum values of both surge and tidal currents are aligned concurrently.

## 2.8 Summary of results and guidelines for application

The propagation of tides generated in ocean basin into estuaries is examined, illustrating the variations in elevation and current responses. The controlling mechanisms are described, explaining how the semi-diurnal and diurnal ocean constituents produce higher-harmonic and residual components within estuaries.

The leading questions are:

> *How do tides in estuaries respond to shape, length, friction factor and river flow?*
> *Why are some tidal constituents amplified yet others reduced and why does this vary*
> *from one estuary to another?*

These questions are addressed by synthesising the system dynamics within simplified equations. This involves linearisation of first-order terms and omission of 'second-order' terms. Omission sets limits of applicability of the derived analytical solutions to estuarine regions where processes are within related scaling bounds. Similarly, linearisation only remains valid within restricted parameter ranges.

In most estuaries, lateral variations in surface elevations are restricted by the large length to breadth ratio; hence, for 'first-order' responses, it is sufficient to concentrate on axial variations. Axial sea level gradients constitute the effective driving force for tidal propagation, and cross-sectionally averaged solutions are appropriate to describe tidal elevations amplitudes, $\varsigma^*$. By contrast, values of the tidal current amplitude, $U^*$, are sensitive to localised variations in both depth and the bed friction coefficient. Hence, values of $U^*$ vary significantly, not only axially but transversely and vertically, Chapter 3 explores these variations more fully.

Proceeding from (2.1) to (2.11), the conditions necessary to reduce the fully 3D non-linear equations to a 1D linearised form are described. The focus is on tidally dominated estuaries, i.e. meso- and macro-tidal estuaries where $\varsigma^*$ at the mouth exceeds 1 m. In such estuaries, the principal lunar semi-diurnal tidal constituent, $M_2$, generally predominates, i.e. has an elevation amplitude at the mouth greater than the sum of all other constituents. This characteristic enables the relevant equations to be linearised directly, in terms of this $M_2$ constituent. Related details of how the quadratic bed friction term is linearised are described in Section 2.5. Similarly, for the axial convective term $U\,\partial U/\partial X$, Section 2.6 illustrates how such a product of velocities at one frequency $\omega$ produces constituents at zero frequency ($Z_o$, residual) and $2\omega$ (or $M_4$ for $\omega = M_2$). The significance of the variation in cross-sectional areas between high and low waters in the generation of both $M_4$ and $Z_0$ constituents is described in Section 2.6.3.

Equation (2.13) provides an explicit solution for $U^*$ in terms of the elevation gradient. This solution simplifies for the contrasting extreme cases of estuaries which are (i) shallow and frictionally dominated (2.14) and (ii) deep and frictionless (2.15). Many earlier studies have focused on the latter. This demarcation between the relative influences of bed friction and the temporal acceleration (inertia) is illustrated in Figs. 2.1 and 2.4. The generalised solution (2.13) indicates how these two terms are orthogonal (i.e. have tidal phases separated by 90°) and together balance the sea level gradient. Figure 2.2 illustrates how this balance holds for $M_2$ but not for $M_4$ or $M_6$ where the effective driving forces is no longer surface gradient but non-linear terms derived from the propagation of $M_2$.

Explicit solutions for $U^*$ in terms of $\varsigma^*$ can be obtained by invoking the 'synchronous' estuary approximation, i.e. where surface gradients due to axial variations in tidal phase are much greater than those due to variations in amplitude. For this approximation, Fig. 2.3 indicates how in most estuaries, tidal velocities generally range from 0.5 to 1.0 m s$^{-1}$. Similarly, for these 'synchronous' solutions, the ratio of the friction to inertial terms approximates $10\,\varsigma^*:D$, where $D$ is the water depth. Further features of this solution, namely phase lag between $\varsigma^*$ and $U^*$, estuarine length and bathymetry, are explored in Chapter 6 while Chapter 7 describes related sorting and trapping of sediments.

Analytical solutions for whole-estuary tidal responses require some functional specification of geometry. Two cases are examined with axial variations in breadths and depths described by (i) $X^n$ and $X^m$ (the synchronous approximation is equivalent to $m=n=0.8$) and (ii) exp $(\alpha X)$ and exp$(\beta X)$. The first of these bathymetric approximations provides a generalised response diagram (Fig. 2.5), showing amplitude and phase variations along any such estuary as a function of the funnelling parameter $v=(n+1)/(2-m)$. Maximum amplification occurs for $v=1$ and nodal lengths, analogous to 'quarter-wave length amplification' in a frictionless prismatic channel, are indicated.

The differences in these responses for a range of tidal constituents and variations in the friction coefficient are illustrated in Fig. 2.6 for the Bay of Fundy. This example illustrates how, typically, diurnals show little amplification, whereas higher harmonics are often greatly amplified.

The corresponding solutions for the second case of exponential representations of bathymetry indicate (Fig. 2.7), how for $\alpha+2\beta<4\pi$, estuarine responses resemble that of an under-damped oscillator while for $\alpha+2\beta>4\pi$ estuaries are over-damped and will show little change in tidal elevation amplitude.

These generalised response Frameworks vary according to the value of the linearised friction factor, illustrating how bathymetry and friction together determine the nature of tidal propagation in estuaries. Moreover, through the adoption of dimensionless parameters, the Frameworks can explain the tidal response at any point, for all tidal constituents, along any funnel-shaped estuary.

Fig. 2.8 illustrates surge–tide interaction in the Thames where the surge and tide components were of roughly equal magnitude and 'periodicities' – effectively precluding linearisation against the $M_2$ tidal constituent. This example emphasises the magnitude and complexity of interactions in such cases.

# References

Aubrey, D.C. and Speer, P.E., 1985. A study of nonlinear tidal propagation in shallow inlet/ estuarine system Part I: Observations. *Estuarine, Coastal and Shelf Science*, **21** (2), 185–205.

Bowden, K.F., 1953. Note on wind drift in a channel in the presence of tidal currents. *Proceedings of the Royal Society of London*, A, **219**, 426–446.

Cartwright, D.E., 1968. A unified analysis of tides and surges round north and east Britain. *Philosophical Transactions of the Royal Society of London*, A, **263** (1134), 1–55.

Dorrestein, R., 1961. *Amplification of Long Waves in Bays*. Engineering progress at University of Florida, Gainesville, **15** (12).

Friedrichs, C.T. and Aubrey, D.G. 1988. Non-linear distortion in shallow well-mixed estuaries; a synthesis. *Estuarine, Coastal and Shelf Science*, **27**, 521–545.

Friedrichs, C.T. and Aubrey, D.G., 1994. Tidal propagation in strongly convergent channels. *Journal of Geophysical Research*, **99** (C2), 3321–3336.

Garrett, C., 1972. Tidal resonance in the Bay of Fundy. *Nature*, **238**, 441–443.

Garrett, C.J.R. and Greenberg, D.A., 1977. Predicting changes in tidal regime: the open boundary problem. *Journal of Physical Oceanography*, **7**, 171–181.

Garrett, C.J.R. and Munk, W.H., 1971. The age of the tides and the Q of the oceans. *Deep Sea Research*, **18**, 493–503.

Gill, A.E., 1982. *Atmosphere-Ocean Dynamics*. Academic Press, New York.

Godin, G., 1988. The resonant period of the Bay of Fundy. *Continental Shelf Research*, **8** (8), 1005–1010.

Heaps, N.S., 1967. Storm surges. In: Barnes, H. (ed.), *Oceanography and Marine Biology Annual Review*, Vol. 5. Allen and Unwin, London, pp. 11–47.

Heaps, N.S., 1983. Storm surges, 1967–1982. *Geophysical Journal of the Royal Astronomical Society*, **74**, 331–376.

Hunt, J.N., 1964. Tidal oscillations in estuaries. *Geophysical Journal of the Royal Astronomical Society*, **8**, 440–455.

Hunter, J.R. 1975. A note on quadratic friction in the presence of tides. *Estuarine, Coastal Marine Science*, **3**, 473–475.

Ianniello, J.P., 1977. Tidally-induced residual currents in estuaries of constant breadth and depth. *Journal of Marine Research*, **35** (4), 755–786.

Ianniello, J.P., 1979. Tidally-induced currents in estuaries of variable breadth and depth. *Journal of Physical Oceanography*, **9** (5), 962–974.

Jeffreys, H., 1970. *The Earth*, 5th edn. Cambridge University Press, Cambridge.

Lamb, H., 1932. *Hydrodynamics*, 6th edn. Cambridge University Press, Cambridge.

Larouche, P., Koutitonsky V.C., Chanut, J.-P., and El-Sabh, M.I., 1987. Lateral stratification and dynamic balance at the Matane transect in the lower Saint Lawrence Estuary. *Estuarine and Coastal Shelf Science*, **24** (6), 859–871.

LeBlond, P.M., 1978. On tidal propagation in shallow rivers. *Journal of Geophysical Research*, **83** (C9), 4717–4721.

Munk, W.H. and Cartwright, D.E., 1966. Tidal spectroscopy and prediction. *Philisophical Transactions of Royal Society of London*, A, **259**, 533–581.

Nihoul, J.C.J. and Ronday, F.C., 1975. The influence of the tidal stress on the residual circulation. *Tellus*, **27**, 484–489.

Prandle, D., 1978. Residual flows and elevations in the southern North Sea. *Proceedings of the Royal Society of London*, A, **359** (1697), 189–228.

Prandle, D., 1980. Modelling of tidal barrier schemes: an analysis of the open-boundary problem by reference to AC circuit theory. *Estuarine and Coastal Marine Science*, **11**, 53–71.

Prandle, D., 1985. Classification of tidal response in estuaries from channel geometry. *Geophysical Journal of the Royal Astronomical Society*, **80** (1), 209–221.

Prandle, D., 2003. Relationship between tidal dynamics and bathymetry in strongly convergent estuaries. *Journal of Physical Oceanography*, **33**, 2738–2750.

Prandle, D., 2004. How tides and river flows determine estuarine bathymetries. *Progress in Oceanography*, **61**, 1–26.

Prandle, D. and J. Wolf., 1978. The interaction of surge and tide in the North Sea and River Thames. *Geophysical Journal of the Royal Astronomical Society*, **55** (1), 203–216.

Prandle, D. and Rahman M., 1980. Tidal response in estuaries. *Journal of Physical Oceanography*, **10** (10), 1552–1573.

Proudman, J., 1923. Report of British Association for the Advancement of Science. *Report of the Committee to Assist Work on Tides*. pp. 299–304.

Saunders, P.H., 1977. Average drag in an oscillatory flow. *Deep Sea Research*, **24**, 381–384.

Taylor, G.I., 1921. Tides in the Bristol Channel. *Proceedings of the Cambridge Philosophical Society/Mathematical and Physical Sciences*, **20**, 320–325.

Van Veen, J., 1947. Analogy between tides and AC electricity. *Engineering*, **184**, 498, 520–544.

Xiu, R. 1983. A study of the propagation of tide wave in a basin with variable cross-section. *First Institute of Oceanography, National Bureau of Oceanography*. Qingdao/Shandong, China.

Zimmerman, J.T.F., 1978. Topographic generation of residual circulation by oscillatory (tidal) currents. *Geophysical and Astrophysical Fluid Dynamics*, **11**, 35–47.

# 3

# Currents

## 3.1 Introduction

The factors determining the magnitudes of depth-averaged tidal currents were described in Chapter 2. Here we explore the vertical structure of both tidal- and wind-driven currents. The structure of density-driven currents is described in Chapter 4. These structures are incorporated into subsequent theories relating to salinity intrusion, Chapter 4; sediment dynamics, Chapter 5 and morphological equilibrium, Chapters 6 and 7.

Models of tidal propagation involve numerical solutions to the momentum and continuity equations. In shelf seas, given adequate numerical resolution, the accuracy of simulations depends primarily on the specification of open-boundary conditions and water depths. Thus, the early 2D (vertically averaged) shelf-sea models (Heaps, 1969) paid scant attention to the specification of bed-stress coefficients. By contrast, applications in estuaries and bays often involve extensive calibration procedures requiring careful adjustment of bed friction coefficients (McDowell and Prandle, 1972). This predominant influence of frictional dissipation in shallow macro-tidal estuaries was illustrated in Chapter 2. In the more-recent 3D models, accurate specification of vertical eddy viscosity, $E$, is similarly essential to reproduce vertical current structure and related temperature and salinity distributions.

Validation of estuarine models of tidal propagation is often limited to comparisons against tide gauge recordings of water levels. In large estuaries with appreciable changes in phase and amplitude, accurate simulation of tidal elevations implies reasonable reproduction of depth-averaged currents (assuming accurate bathymetry). However, in small estuaries, tidal elevations often show little variation in either phase or amplitude from the open-boundary specifications, and so such validation offers little guarantee of accurate representation of tidal currents.

In comparison with tidal elevations, currents are characterised by significant variability both temporally and spatially. Thus, while observations of water levels in estuaries typically have a (non-tidal component) noise: tidal signal ratio of O (0.2), for currents these components are often of equal magnitudes. *In situ* current measurements can be made via mechanical, electro-magnetic or acoustic sensors, while surface measurements can be made remotely using H.F. Radar. Such measurements are generally more costly, less accurate and less representative than for elevations. The spatial inhomogeneity of currents complicates the use of adjacent observations in prescribing 'related constituents' for tidal analyses of short-term recordings. Hence, extended observational periods are necessary to accurately separate constituents, especially at the surface where contributions from wind- and wave-driven current components are largest.

Since the tidal solutions described in Sections 3.2 and 3.3 omit wind forcing, density and convective terms, a brief description of their relative magnitudes is presented. While the following description provides simple scaling analyses, the presence of significant wind stress or density gradients may radically change the magnitude and vertical distribution of the vertical eddy viscosity coefficient and thus, interactively, the tidal current structure. Souza and Simpson (1996) provide a good example of how the vertical structure of tidal current ellipses can be radically changed by pronounced stratification.

The depth-averaged axial momentum equation may be written as

$$
\begin{aligned}
\frac{\mathrm{d}U}{\mathrm{d}t} &= \frac{\partial U}{\partial t} + U\frac{\partial U}{\partial X} + V\frac{\partial U}{\partial Y} \\
&= -g\frac{\partial \varsigma}{\partial X} - 0.5D\frac{\partial \rho}{\rho \partial X} - \frac{fU(U^2 + V^2)^{1/2}}{D} - \frac{f_{\mathrm{w}}W^2}{D} + \Omega V,
\end{aligned}
\tag{3.1}
$$

where $U$ and $V$ are velocities along axial and transverse horizontal axes $X$ and $Y$; $W$ is the wind speed; $\varsigma$ is surface elevation; $\rho$ density; $D$ water depth; $\Omega$ the Coroilis parameter; $f$ and $f_{\mathrm{W}}$ coefficients linked with bed friction and wind drag, respectively.

### 3.1.1 Convective term

It was shown in Section 2.2 that the ratio of the inertial to the axial convective term can be approximated by

$$
\frac{\partial U}{\partial t} : U\frac{\partial U}{\partial X} \sim c : U,
\tag{3.2}
$$

where $c = (gD)^{1/2}$ is the wave celerity. Tidal currents only approach 'critical' celerity, $c = U$, in confined sections, suggesting that omission of the convective terms is generally valid. However, non-linear coupling of tidal constituents introduced by this term may often be significant for higher-harmonic constituents, as described in Section 2.6.

The transverse convective term $V\,\partial U/\partial Y$ can be significant close to coastline features and in regions of sharp changes in bathymetry (Pingree and Maddock, 1980; Zimmerman, 1978). Prandle and Ryder (1989) present detailed quantitative analyses of the roles of convective terms in a fine-resolution numerical model simulation. The spatial signatures associated with these terms have been mapped both by H.F. Radar (Fig. 3.9; Prandle and Player, 1993) and from ADCP current observations (Geyer and Signell, 1991).

### 3.1.2  Density gradients

The influence on current structure of axial density gradients and vertical stratification associated with saline intrusion is described in Chapter 4. Density stratification associated with heat exchange at the water surface may be significant when the (bed) frictional boundary layer does not extend through the whole depth, (Appendix 4A). This is generally confined to the deepest, micro-tidal estuaries.

From (3.1), the ratios of the surface gradients associated with elevation and saline intrusion are in the ratio

$$\frac{2\pi\varsigma}{\lambda} : 0.5D\frac{0.03}{L_I} \qquad \text{or} \qquad \frac{\varsigma}{D} : \frac{0.002\lambda}{L_I}, \tag{3.3}$$

where $L_I$ is the salinity intrusion length and $0.03\rho$ is the additional density of sea water.

Hence, in terms of surface elevation, the saline density gradient will only be important in micro-tidal, deep estuaries. While salinity intrusion can significantly change the vertical structure of currents, Prandle (2004) confirmed that it has little impact on tidal levels in estuaries.

### 3.1.3  Wind forcing

The wind-induced stress at the sea surface may be approximated by (Flather, 1984)

$$\tau_w = 0.0013\,W^2\,\mathrm{nm}^{-2}. \tag{3.4}$$

The equivalent bed-stress term for tidal current $U$ is

$$\tau_B = 0.0025\rho\,U^2. \tag{3.5}$$

Thus, for wind stress to exceed the bed stress, the wind speed $W > 44U$.

### 3.1.4  Approach

In Section 3.2, we assume that flow is confined to the axial direction, $X$. Then, specifying a surface gradient, the effects on tidal current structure of the bed friction coefficient, $f$, and vertical eddy viscosity, $E$, are examined. These solutions are valid in long narrow estuaries, i.e. $\Omega B \ll c$ ($B$, breadth; Ianniello, 1977). In such estuaries, little transverse variation can occur in the axial component of surface gradients and cross-sectional variations in currents correlate directly with depth variations (Lane *et al.*, 1997). This section derives the scaling laws that explain the diversity of tidal current structure.

In wider estuaries, flow is more 3-dimensional, and, in Section 3.3, the impacts of the Earth's rotation via the Coriolis force are demonstrated. Since the Coriolis force depends directly on latitude, the acute sensitivity of tidal currents close to the inertial latitudes ($\sim 30°$ for diurnal and $\sim 70°$ for semi-diurnal) is shown.

Section 3.4 examines time-averaged tidal- and wind-driven currents. Whereas for elevations, these components are generally small in relation to the predominant oscillatory tidal variations, for currents these components can be comparable, especially during extreme events.

## 3.2 Tidal current structure – 2D (*X-Z*)

In this section, it is assumed that tidal currents are confined to rectilinear flow along the $X$-axis and can be determined from the linearised equations of motion. Theories of the vertical structure of tidal- and wind-driven currents remain largely dependent on the concept of a vertical eddy viscosity parameter, $E$. The analytical solutions derived here (Prandle, 1982a) assume a constant value of $E$. Prandle (1982b) extended the present derivation to the case of $E$ varying linearly with height above the bed.

Throughout this chapter, the 'bed' can be regarded as the interface between the logarithmic layer and the Ekman layer (Bowden, 1978). Precise simulation of vertical structure closer ($\sim 1$ m) to the bed involves consideration of boundary layer theory and variations of $E$ with time and depth. Such simulations must ensure that this logarithmic velocity profile is smoothly matched with the exterior flow (Liu *et al.*, 2008). Section 3.3 includes comparisons between the present analytical solutions and the currents derived from detailed numerical simulations in which $E$ is calculated from a 'turbulent energy closure model' described in Appendix 3B.

### *3.2.1 2D analytical solution*

The momentum equation for tidal flow confined to one horizontal dimension, neglecting vertical acceleration, wind forcing, convective and density terms, can be expressed as

$$\frac{\partial U}{\partial t} = -g \frac{\partial \varsigma}{\partial X} + \frac{1}{\rho} \frac{\partial}{\partial Z} F_z, \qquad (3.6)$$

where $F_z$ is the component of frictional stress exerted at level $Z$ by the water above that level. Expressing the frictional stress in terms of a (constant) vertical eddy viscosity, $E$, gives

$$F_z = \rho\, E\, \frac{\partial U}{\partial Z}. \tag{3.7}$$

Limiting consideration to a single tidal constituent of frequency, $\omega$, at any position, we assume

$$U(z,t) = \text{Re}[U(Z)e^{i\omega t}] \tag{3.8}$$

and

$$\zeta(t) = \text{Re}[We^{i\omega t}], \tag{3.9}$$

where $U(Z)$ and $W$ take a complex form to reflect tidal phase variations.

Substituting (3.7), (3.8) and (3.9) into (3.6), we can eliminate the time variation $e^{i\omega t}$ to give

$$i\,\omega\, U = -g\frac{\partial}{\partial X}\, W + E\, \frac{\partial^2}{\partial Z^2}\, U. \tag{3.10}$$

Equation (3.10) is satisfied by the solution

$$U = A_1 e^{bz} + A_2 e^{-bz} + C \tag{3.11}$$

with

$$b = \left(\frac{i\omega}{E}\right)^{1/2} \text{and } C = \frac{-g}{i\omega}\frac{\partial}{\partial X}\, W. \tag{3.12}$$

*Boundary conditions*

At the surface $Z=D$, the frictional stress $F_z=0$, i.e.

$$A_1 b\, e^{bD} - A_2 b\, e^{-bD} = 0$$

$$\text{or} \quad A_1 = A_2\, e^{-2bD}. \tag{3.13}$$

At the bed $Z=0$, we assume the stress described by (3.7) is equal to the stress described by the linearised quadratic friction law, Section 2.5 (Proudman, 1953, p. 316),

$$E\,(A_1 b - A_2 b) = \frac{8}{3\pi}\, f\, U^*\, U_{z=0} = \frac{8}{3\pi}\, f\, U^*\,(A_1 + A_2 + C), \tag{3.14}$$

where $U^*$ represents the depth-averaged tidal amplitude.

*Continuity*

The total flow is given by

$$U^* D = \int_0^D U \, dZ = \frac{A_1}{b} \left(e^{bD} - 1\right) - \frac{A_2}{b} \left(e^{-bD} - 1\right) + CD. \qquad (3.15)$$

*Velocity profiles*

Combining (3.11) to (3.15), we obtain the following solution for the velocity $U$ at any height $Z$,

$$\frac{U(z)}{U^*} = \frac{(e^{bZ} + e^{-bZ+2y})}{T} + Q, \qquad (3.16)$$

$$\text{where} \quad T = (1 - e^{2y})\left(\frac{j-1}{y-1}\right) - 2\,e^{2y} \qquad (3.17)$$

$$\text{and} \quad Q = \frac{[j(1 - e^{2y}) - 1 - e^{2y}]}{T} \qquad (3.18)$$

$$j = \frac{3\,\pi\,E\,b}{8\,f|U^*|} \qquad (3.19)$$

$$y = bD. \qquad (3.20)$$

Letting $j = J\,i^{1/2}$ and $y = Y\,i^{1/2}$, we obtain

$$J = \frac{3\pi(E\omega)^{1/2}}{8\,f\,U^*} \qquad (3.21)$$

$$\text{and} \quad Y = \left(\frac{\omega}{E}\right)^{1/2} D. \qquad (3.22)$$

### 3.2.2  Comparisons with observations

The velocity profile described by (3.16) is a function of two variables, $J$ and $Y$. $Y$ may be interpreted as a depth parameter converted to a dimensionless form by Ekman scaling (Faller and Kaylor, 1969; Munk *et al.*, 1970). $J$ is also dimensionless and reflects the effect of the quadratic bottom stress through the bed-stress coefficient $f$ and the depth-averaged velocity amplitude $U^*$. It may be shown from (3.16) that when $J$ is large, the velocity distribution is always uniform through depth. For large values of $Y$, the solution (3.16) approaches an asymptote that is a function

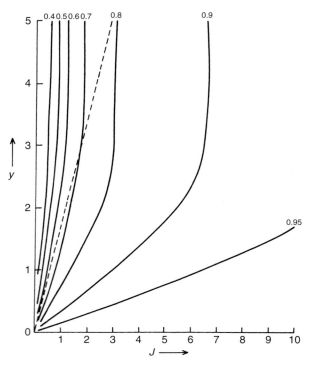

Fig. 3.1. Surface to bed differences in amplitude and phase of tidal currents as *f* (*J, Y*). (Left) Velocity at the bed as a fraction of the depth-averaged value. (Right) Phase differences between velocities at the surface and bed (negative values indicate phase advance at the bed).

of *J* only. Thus, the full range of vertical tidal current structure is represented by the parameter ranges $0 < Y < 5$ and $0 < J < 10$.

Figure 3.1 shows, in contour form, the ratio of the velocity at the bed to the depth-averaged velocity, i.e. $|U_{Z=0}/U^*|$ plotted as a function of $Y$ and $J$. It also shows the phase difference $\Delta\theta = \theta_s - \theta_b$ between velocities at the surface and bed, with $\Delta\theta < 0$ indicating a phase advance at the bed. Pronounced vertical structure occurs in the range $Y > 1$ and $J < 2$. This latter point is emphasised by Fig. 3.2 (Prandle, 1982a) where complete velocity profiles are shown for co-ordinate values, $(J, Y) = $ (a) (0.5,5); (b) (5,5); (c) (0.5,0.5) and (d) (5,0.5).

The profiles described by the above theory correspond in character with measured profiles for tides in rivers, estuaries and shallow seas (Van Veen, 1938). The depth mean velocity occurs at a fractional height $(z = Z/D)$, $0.25 < z < 0.42$ – validating the common engineering assumption that average velocity occurs close to $0.4D$ above the bed. Velocities measured at $z = 0.4D$ will provide an estimate of the depth mean value with a maximum error of 4% in amplitude and 1.5° in phase.

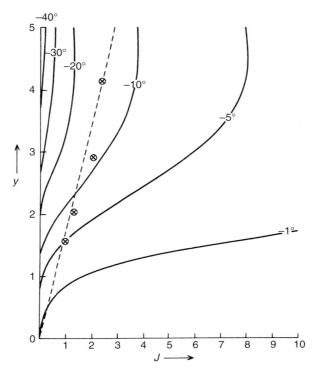

Fig. 3.1. (cont.)

The maximum velocity generally occurs close to the surface. However, for $Y > 4$ maximum velocity occurs below the surface, and for $Y \sim 5$ maximum velocity occurs at mid-depth.

### 3.2.3 Eddy viscosity formulation, Strouhal number

A fuller interpretation of the above theory requires some formulation for eddy viscosity. Two commonly assumed expressions in dimensionally consistent form are

$$E = \alpha U^* D \tag{3.23}$$

$$E = \frac{\beta U^{*2}}{\omega}. \tag{3.24}$$

Bowden (1953) suggested a formulation similar to (3.23) while Kraav (1969) suggested a formulation similar to (3.24). Substituting (3.23) into (3.21) and (3.22) we obtain

$$Y = \frac{8}{3\pi} f \frac{J}{\alpha}. \tag{3.25}$$

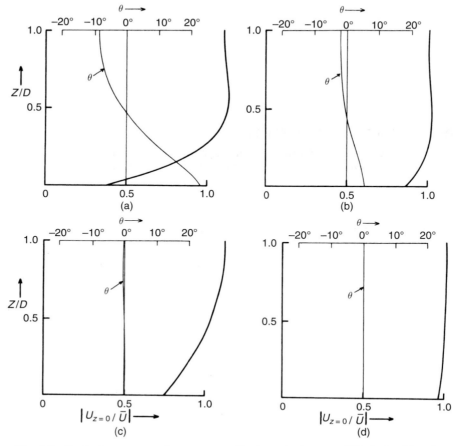

Fig. 3.2. Vertical structure of current amplitude and phase as $f(J, Y)$. Profiles for $(J, Y) =$ (a) $(0.5, 5.0)$, (b) $(5.0, 5.0)$, (c) $(0.5, 0.5)$ and (d) $(5.0, 0.5)$.

Similarly, substituting (3.24) into (3.21) gives

$$J = \frac{\beta^{1/2}}{8/3\pi} f. \tag{3.26}$$

McDowell (1966) presents values of the phase difference $\theta$ between surface and bed found for oscillatory flows in a laboratory flume, showing that $\theta$ increases continuously as the parameter $f S_R$ decreased, where $S_R = U P/D$ is the Strouhal number. Prandle (1982a) showed that these data correspond to $Y = 1.7J$, thus supporting Bowden's (1953) formulation for eddy viscosity and from (3.25) suggesting a value of $\alpha \sim 0.5f$. Using $E = \alpha' \mid U_{z=0} \mid D$, Bowden found values of $\alpha'$ in the range 0.0025–0.0030, thus agreement with the present value for $\alpha$ requires

$|U_{z=0}/U| \sim 0.5$. Prandle (1982b) showed that analytical solutions of (3.6) require $E > 0.5 f U^* D$.

Bowden's (1953) formulation is appropriate in most estuaries when the boundary layer thickness extends to the surface, whereas in deeper water, Kraav's (1969) formulation applies. Davies and Furnes (1980) used the latter formulation in modelling tidal current structure in depths of up to 200 m over the UK continental shelf area.

By assuming Bowden's (1953) formulation for eddy viscosity, with $\alpha = f$, the vertical structure (3.16) reduces to a function of the single parameter $f S_R$, with $Y \sim 0.83 J \sim 50/S_R^{1/2}$. Figure 3.3 (Prandle, 1982a) shows the resultant vertical structure as a function of $S_R$ for the case of $f = 0.0025$. For small values of $S_R$, i.e. $S_R < 50$, the current structure is uniform except for a small phase advance close to the bed. For larger values of $S_R$, the variation in current amplitude increases continuously, approaching an asymptote in the region of $S_R = 1000$. The phase variation also increases with increasing $S_R$ but reaches a maximum difference between surface and bed in the region of $S_R = 350$. Thereafter, the phase variation decreases with increasing $S_R$ with only 1° difference between surface and bed for $S_R = 10\,000$. The phase variation is generally concentrated close to the bed except when $100 < S_R < 1000$.

In a typical strongly tidal estuary with an $M_2$ tidal current amplitude $U^* \sim 1\ \mathrm{ms}^{-1}$, $S_R > 1000$ for $D < 40$ m, and thus the vertical structure will tend towards the asymptotic solution for large $S_R$ shown in Fig. 3.3.

## 3.3 Tidal current structure – 3D (*X-Y-Z*)

The fundamental difference in extending the above theory for uni-directional to fully 3D currents is the inclusion of the Coriolis force in the momentum equation (3.6). Although for the major tidal constituents in many estuaries, the depth-integrated effect of this term may be of second order, the (differential) influence on vertical structure can be highly significant.

### 3.3.1 Tidal ellipse resolved into clockwise and anti-clockwise components

In a horizontal plane, tidal current vectors rotate and, relative to a fixed origin, describe an elliptical path. A common practice is to resolve the ellipse into two circular motions, one rotating clockwise ($R_2$, $g_2$) and the other anti-clockwise ($R_1$, $g_1$). Appendix 3A provides a fuller description of this transformation. In vector notation with $X$ the real axis and $Y$ the orthogonal complex axis, the tidal vector, $\boldsymbol{R}$, is given by

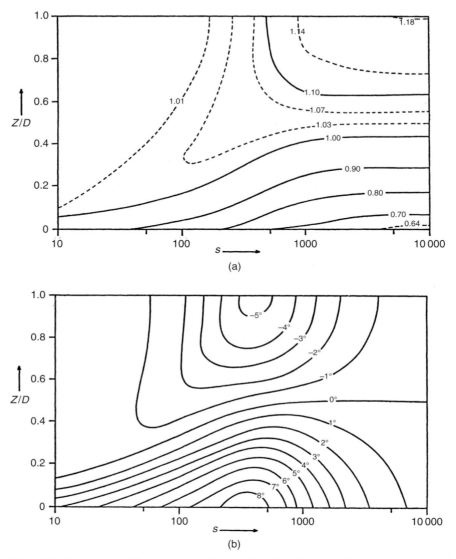

Fig. 3.3. Current profile as a function of the Strouhal number, $S_R = U^* P/D$. (a) amplitude $U(z)/U_{mean}$; (b) phase structure $\theta(z) - \theta_{mean}$.

$$R = U + iV. \tag{3.27}$$

Resolving $R$ into a clockwise component $R_2$ and an anti-clockwise component $R_1$ ($|R_1|$ and $|R_2|$ constant),

$$R = R_1 + R_2. \tag{3.28}$$

### 3.3.2  3D analytical solution

The linearised equations of motion for flow in two dimensions may be written as

$$\frac{\partial U}{\partial t} - \Omega V = -g\frac{\partial \varsigma}{\partial X} + \frac{1}{\rho}\frac{\partial}{\partial Z}F_{zx} \qquad (3.29)$$

and

$$\frac{\partial V}{\partial t} + \Omega U = -g\frac{\partial \varsigma}{\partial Y} + \frac{1}{\rho}\frac{\partial}{\partial Z}F_{zy}, \qquad (3.30)$$

where $\Omega$ is the Coriolis parameter and $F_{zx}$ and $F_{zy}$ are the components of $F_z$ along $X$ and $Y$. These equations can be transformed using (3.27) and (3.28) into the following equations for the separate rotational components:

anti-clockwise $\qquad i\,(\Omega + \omega)\ \boldsymbol{R}_1 = \boldsymbol{G}_1 + \dfrac{\partial}{\partial Z}\,E\,\dfrac{\partial}{\partial Z}\,\boldsymbol{R}_1 \qquad (3.31)$

clockwise $\qquad i\,(\Omega - \omega)\ \boldsymbol{R}_2 = \boldsymbol{G}_2 + \dfrac{\partial}{\partial Z}\,E\,\dfrac{\partial}{\partial Z}\,\boldsymbol{R}_2, \qquad (3.32)$

where $\boldsymbol{G}_1$ and $\boldsymbol{G}_2$ are rotational components of the surface gradient.

Comparing (3.31) and (3.32) with (3.10) we see that for $E$ constant through depth, the equations are analogous. Thus, the solutions will be equivalent with the important difference that for the anti-clockwise component

$$b \to B_1 = \left[\frac{i\,(\Omega + \omega)}{E}\right]^{1/2}, \qquad (3.33)$$

while for the clockwise component

$$b \to B_2 = \left[\frac{i\,(\Omega - \omega)}{E}\right]^{1/2}. \qquad (3.34)$$

For $\Omega > \omega$, the resulting velocity profiles can be deduced directly from the 1D profiles described previously by simply replacing $\omega$ by $\omega' = \Omega \pm \omega$. For $\Omega < \omega$, the clockwise parameter $B_2$ may be rewritten in the form

$$B_2' = i\left[\frac{i\,(\omega - \Omega)}{E}\right]^{1/2}. \qquad (3.35)$$

From (3.11), velocity structure is dependent on $\exp(\pm bZ)$ then noting that

$$e^{i^{1/2}}q = (e^{iq^2})^{1/2}, \qquad (3.36)$$

whereas

$$e^{i3/2q} = (e^{-iq^2})^{1/2} \qquad (3.37)$$

with $q$ a real constant, we see that the introduction of an additional i in (3.37) simply changes the direction of rotation. In consequence, the sign of the phase difference between surface and bed changes; however, since the vector is rotating in the opposite direction, there remains a phase advance at the bed.

The current structure for the a–c and c–w components of the tidal ellipse can be estimated from Fig. 3.3 by calculating their respective Strouhal numbers as follows:

$$S_{a-c} = \frac{2\pi|R_1|}{(D(\Omega + \omega))} \text{ and } S_{c-w} = \frac{2\pi|R_2|}{(D(\omega - \Omega))}. \qquad (3.38)$$

Thus, for semi-diurnal constituents in latitudes less than 70°, vertical structure will be greater for the clockwise component than for the anti-clockwise component. The more pronounced current structure for the clockwise component means that the tidal current ellipse becomes more positively eccentric towards the bed. (Positive eccentricity indicates that $|R_1| > |R_2|$.) Similarly, it can be deduced that the direction of the major axis of the ellipse will veer in a clockwise sense towards the bed.

At mid-latitudes, for the other major tidal frequency bands, i.e. diurnal, quarter-diurnal, the ratio $(\Omega + \omega) : |\Omega - \omega|$ is smaller than for the semi-diurnal band. Hence, the difference between the velocity structures for the two rotational components should be less than that for $M_2$.

### 3.3.3 Sensitivity to friction factor and eddy viscosity

Appendix 3A indicates how the ellipse parameters – $A$ amplitude of the major axis, $E_C$ eccentricity, $\psi$ direction and $\phi$ phase – can be calculated from the a–c and c–w vector components. In summary, the anti-clockwise current vector $(R_1, \theta_1,)$ and clockwise vector $(R_2, \theta_2)$ are related to the more conventional parameters as follows:

$$A = R_1 + R_2, \quad E_C = \frac{R_1 - R_2}{R_1 + R_2}, \quad \psi = \frac{\theta_2 + \theta_1}{2} \text{ and}$$
$$\phi = \frac{\theta_2 - \theta_1}{2}. \qquad (3.39)$$

Figure 3.4 (Prandle, 1982b) illustrates typical current structures for $M_2$ at latitude 55° N together with the sensitivity to both $f$ and $E$. These results show how, in the vicinity of the bed, reducing $E$ enhances vertical current structure, decreases amplitude, increases eccentricity (in a positive a–c sense) and advances phase. These trends are similar to those shown for increasing bottom friction, but in the latter case, there is an additional reduction in the overall current amplitudes (relative to the frictionless values used in prescribing the surface gradients).

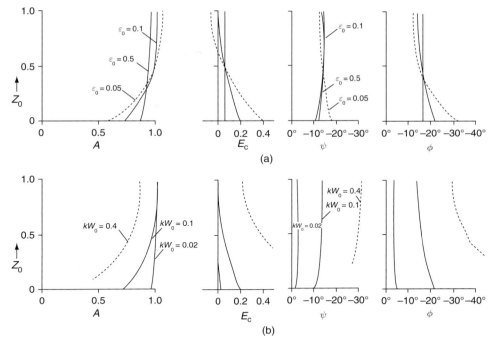

Fig. 3.4. Sensitivity of current structure to eddy viscosity and bed friction.
$A$ amplitude, $E_c$ eccentricity, $\psi$ direction and $\phi$ phase.
(a) Eddy viscosity $E = \varepsilon_0 D^2 \Omega$, $\varepsilon_0 = 0.05$, $0.1$ and $0.5$ ($kW_0 = 0.1$).
(b) Bed stress $\tau = \rho f U^* U$, $f U^* = k W_0 \Omega D$, $kW_0 = 0.02$, $0.1$ and $0.4$ ($\varepsilon_0 = 0.1$).
Results for $M_2$ amplitude of $1 \, \text{m s}^{-1}$ at latitude $55° \, \text{N}$.

### 3.3.4 Sensitivity to latitude

Figure 3.5 (Prandle, 1997) shows solutions for (3.31) and (3.32) as functions of latitude and water depth for surface gradients which correspond to a 'free stream' rectilinear tidal current $R^* = 0.32 \, \text{m s}^{-1}$. (Free stream corresponds to the solution of (3.31) and (3.32) with $D$ infinite or $f$ zero, i.e. in deep, frictionless conditions.) The figure emphasises the enhanced influence of the friction term at latitudes corresponding to the inertial frequency, i.e. for $M_2$: $\sin^{-1}(24/2/12.42) \sim 75°$.

Figure 3.6 (Prandle, 1997) extends these results using simulations by Prandle (1997), from (A) a 2D vertically averaged model and (B) a 3D model incorporating a Mellor and Yamada (1974) level 2.5 closure scheme as described in Appendix 3B. The simulations cover a range of $R^* = 0.1$, $0.32$ and $1.0 \, \text{m s}^{-1}$. The contour values shown are restricted to $|R| = 0.9R^*$, phase and direction $= \pm 10°$ and eccentricity $= \pm 0.1$. These can be regarded as representative boundaries between conditions where tidal propagation is little influenced by friction (in deeper water) and conditions where bottom friction becomes increasingly significant (in shallower water).

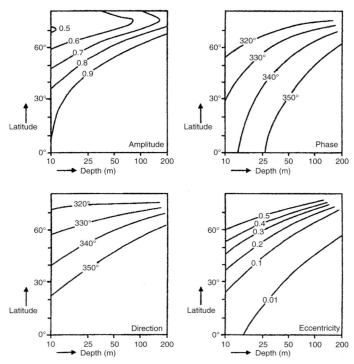

Fig. 3.5. Modulation of depth-averaged $M_2$ ellipse due to bed friction, as $f$ (Depth, Latitude). Contours show changes (fractional for amplitude $R = 0.32$ m s$^{-1}$) relative to frictionless values of zero eccentricity, phase and direction.

The qualitative agreement for depth-averaged ellipse parameters between the 2D and 3D models indicates how in deeper water these results are very close, while in shallow water, by suitable adjustment of the bed friction coefficient, 2D model results can be adjusted to approximate those of 3D models. It was shown in Section 2.5 that for a predominant tidal constituent, $i$, the quadratic friction term $f R_i R_i|$ can be linearised (in one dimension) to $(8/3\pi) f R_i R_i^*$ (where $R_i^*$ is the tidal amplitude). The equivalent linearisation for other constituents, j, *is* $(4/\pi) f R_j R_i^*$. Thus, the frictional effect is linearly proportional to $f R_i^*$ for all constituents, and thus any enhancement of $f$ to improve reproduction of the primary constituent in a 2D model should not adversely affect other constituents. The results shown in Fig. 3.6 for $f = 0.0125$ are included as an illustration of an enhanced frictional coefficient applicable for a secondary constituent factored by $3/2 R_i^*/R_j^*$ in such simulations.

### 3.3.5 Surface to bed changes in tidal ellipses

Figure 3.7 (Prandle, 1997) indicates the changes in current ellipse parameters between the surface and the bed as calculated from models: (B) the level 2.5 closure $k$–$\varepsilon$ model

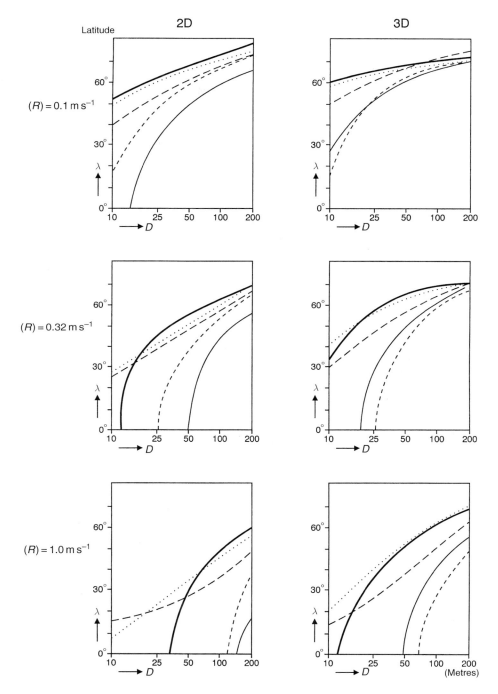

Fig. 3.6. Modulation of depth-averaged $M_2$ ellipse parameters by friction in 2D and 3D models.
(Left) Bed friction in a 2D model;
(Right) bed friction and vertical eddy viscosity (MY 2.5) in a 3D model.
Contours show: — amplitude 0.9, - - - - phase 350°, ———— direction 350°, ……..
eccentricity 0.1 and — bed friction coefficient $f \times 5 = (0.0125)$. For frictionless $R^* = 0.1$
(top), 0.32 (middle), 1.0 (bottom) m s$^{-1}$.

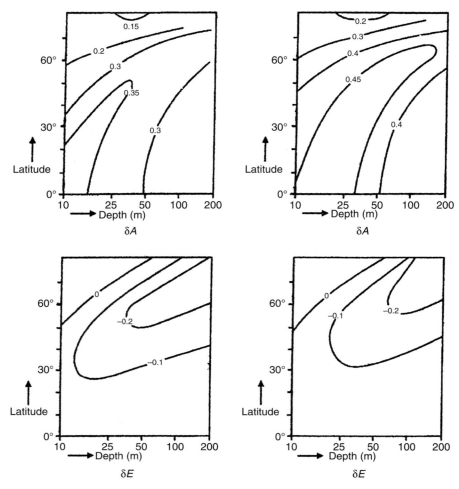

Fig. 3.7. Surface to bed changes in $M_2$ amplitude and eccentricity as $f$ (Depth, Latitude) (amplitude as a fraction of frictionless depth-averaged $R* = 0.32\,\mathrm{m\,s^{-1}}$, $Ec = 0$). (Left) Analytical solution (Section 3.2); (right) 3D numerical model (MY 2.5).

and (C) an analytical solution with $E$ constant (Prandle, 1982a). The close level of agreement shown between these two approaches results from specifying $E \sim E_0$ in the analytical model, where $E_0$ is the time-averaged value of vertical eddy viscosity at the bed computed in the $k$–$\varepsilon$ model. While it can be shown that such tuning can be used to adjust any specific ellipse parameter, it is not possible to force precise agreement for all four parameters simultaneously. Thus, if a detailed representation of current profiles is required, it is necessary to use model (B) to provide a detailed description of the temporal and vertical variations in $E$.

The mean depth-averaged value of $E$ calculated from the $k$–$\varepsilon$ model (B) is typically four times larger than the value at the bed $E_0$, and thus specifying this in model (C) significantly reduces the vertical structure shown in Fig. 3.7.

### *3.3.6 Currents across the inter-tidal zone*

Successive wetting and drying of inter-tidal areas can generate significant cross-shore currents. If it is assumed that such currents are the primary source of in-filling and draining, then tidal current speeds over a uniform bank slope $SL$ will have an amplitude of $\omega\varsigma^*/SL$, where $\varsigma^*$ is the amplitude of the surface elevation. For a semi-diurnal current, this corresponds to a maximum cross-shore current amplitude of $7\,\mathrm{cm\,s^{-1}}$ for every kilometre of exposed bank. Prandle (1991), shows, from both modelling simulations and H.F. Radar observations, how such currents over inter-tidal zones modify tidal ellipse characteristics in the near-shore zone.

## 3.4 Residual currents

Removal of the oscillatory tidal component from current observations leaves residuals that may include contributions from 'rectified' tidal propagation, direct (localised) wind forcing, indirect (larger scale) wind forcing, surface waves and horizontal and vertical density gradients. Interaction of any of these components with the tidal component can generate significant modulation of the latter, contributing an additional non-tidal residual. Selective filtering of the non-tidal currents into frequency bands can be used to separate many of these components.

Density-driven residual currents are described in Chapter 4. Soulsby *et al.* (1993) provide a review of the nature and impact of current interaction with surface waves, quantifying the effect on the bed friction coefficient. Wolf and Prandle (1999) indicate how, in shallow water, such wave impacts can substantially reduce tidal currents.

### *3.4.1 Wind-driven currents*

Relating observed wind-driven currents to wind forcing is notoriously difficult. Both the wind itself and the associated currents exhibit appreciable fine-scale (temporal and spatial) variability. In shallower water, wind forcing may be partially balanced by surface slopes. In constricted cross sections, these slopes can subsequently generate currents which are orders of magnitude greater than indicated by direct localised surface wind forcing (Prandle and Player, 1993).

The response of the sea surface to surface wind forcing under steady-state conditions was originally studied by Ekman; a convenient summary of results is given by Defant (1961). The essential features can be described by considering a steady-state solution to (3.31) and (3.32) (Prandle and Matthews, 1990). Thus rewriting these equations in the form

$$i\Omega R + S = E \frac{\partial^2 R}{\partial Z^2}, \tag{3.40}$$

where $S$ is the surface gradient term and $E$ the vertical eddy viscosity is assumed constant.

Surface ($Z=D$) and bed ($Z=0$) boundary conditions are, respectively,

$$\tau_w = \rho E \frac{\partial R}{\partial Z} \tag{3.41}$$

$$\rho F R_{z=0} = \rho E \frac{\partial R}{\partial Z}, \tag{3.42}$$

where $\tau_w$ is the surface wind stress and $F R_{z=0}$ a (linearised) representation of bed friction.

A solution in the form of an Ekman spiral,

$$R = A e^{bz} + C, \tag{3.43}$$

is satisfied by

$$R(z) = \frac{\tau_w}{\rho E b e^{bD}} \left( e^{bz} + \frac{Eb}{F} - 1 \right), \tag{3.44}$$

where $b^2 = i (\Omega/E)$.

In deeper water, $bD \gg 1$, i.e. $D \gg (E/\Omega)^{1/2}$, the first term in (3.44) predominates and

$$R_{z=D} = \frac{-\tau_w i^{1/2}}{\rho (\Omega E)^{1/2}}, \tag{3.45}$$

i.e. a surface current of magnitude dependent on latitude and veering at 45° clockwise to the wind stress.

In shallow water, the second term predominates and

$$R = \frac{\tau_w e^{-bD}}{\rho F}, \tag{3.46}$$

i.e. a current of magnitude dependent on the bed friction coefficient and aligned with the wind.

Figure 3.8 shows these wind-driven surface-current responses derived from statistical analyses of H.F. Radar observations (Prandle and Matthews, 1990).

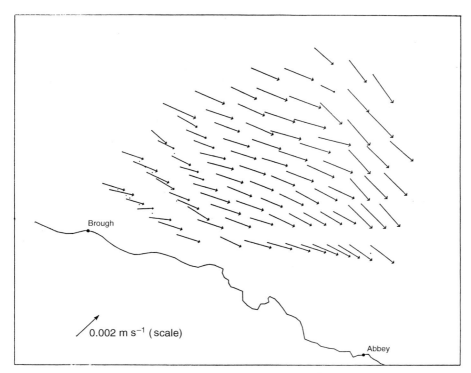

Fig. 3.8. Wind-driven surface current for a wind of 1 m s$^{-1}$ eastwards. Derived from H.F. Radar measurements.

Substituting (3.4) for wind stress, $\tau$, yields a response in close agreement with that shown from the Radar observations in Fig. 3.8. The steady-state surface currents are typically 1 or 2% of wind speed, increasing in deeper water in both magnitude and veering towards the theoretical deep water values of 45° to the right of the wind. The observed veering ranges from 3° to 35° (clockwise).

### 3.4.2 Tidal current residuals

The mechanisms by which tidal energy propagating into estuaries at semi-diurnal and diurnal periods generate higher-order harmonics and residual components is described in Section 2.6. It was shown how determination of residual currents is complicated by which parameter is defined to be the (linear) controlling factor. In shallow water, calculation of residual current components is further complicated by the respective reference systems, e.g. fixed distances above the bed or below the surface, fractional heights. The continuous profile provided by an Acoustic Doppler Current Profiler (ADCP) allows observational data to be readily interpolated onto the selected vertical framework. However, for residual and higher harmonics, derived values are especially sensitive to the prescribed framework (Lane *et al.*, 1997).

An Euler current at a fixed height above the bed corresponds to the velocity observed by a moored current meter. The tidally averaged, depth-integrated Euler residual is defined as

$$U_E = \frac{1}{P} \int_0^P \left( \frac{1}{D+\varsigma} \int_0^{D+\varsigma} U(Z)\, dZ \right) dt. \tag{3.47}$$

The tidally averaged, depth-integrated, residual transport current is

$$U_T = \frac{1}{DP} \int_0^P \left( \int_0^{D+\varsigma} U(Z)\, dZ \right) dt \tag{3.48}$$

with $U_T D = Q$, the river flow per unit breadth.

For uniform flow of a single tidal constituent in one direction, the difference between the above residual current components is

$$U_T = U_E + 0.5\,\varsigma^*\, U^* / D \cos(\theta), \tag{3.49}$$

where $\theta$ is the phase difference between $\zeta$ and $U$ and the second term is referred to as the Stokes' drift. Cheng *et al.* (1986) provide further details of the difference between Eulerian and Lagrangian flows in non-uniform flow fields.

Prandle (1975) showed that the depth-integrated net energy propagation can be approximated by

$$EN = \rho g D [0.5\,\varsigma^*\, U^* \cos(\theta)]. \tag{3.50}$$

This implies that net propagation of tidal energy will be accompanied by a small residual Stokes' drift. In open seas, this can produce a persistent residual circulation. A modelling study by Prandle (1984) showed typical residual currents of $1\text{–}3\ \mathrm{cm\,s^{-1}}$ in the continental shelf around the UK. However, in an enclosed estuary, some compensating outflow must be present, as noted in Section 2.6.3.

As an example of these persistent tidally driven residual currents, Fig. 3.9 shows a residual surface current gyre measured in a year-long H.F. Radar deployment in the Dover Strait (Prandle and Player, 1993). Such gyres are not exceptional but rather a characteristic of most coastlines, although rarely identified with conventional instruments. Geyer and Signell (1991) mapped a similar headland gyre with a vessel-mounted ADCP.

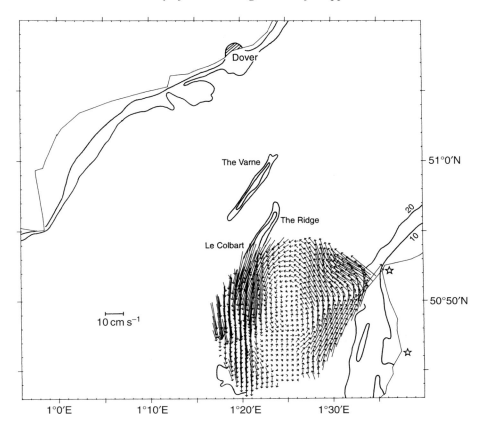

Fig. 3.9. Residual tidal currents observed at the surface by H.F. Radar.

## 3.5 Summary of results and guidelines for application

Despite large ebb and flood tidal excursions within estuaries, longer-term mixing of salt, sediments and pollutants can remain sensitive to small-scale variability in currents and persistent residual circulation. Hence, the focus here is on deriving the scaling factors which determine the vertical structure of currents, illustrating the sensitivity to tidal current amplitude, tidal period, depth, friction factor and latitude.

The key question is:

*How do tidal currents vary with depth, friction, latitude and tidal period?*

Equation (2.13) shows how the depth-averaged tidal velocity amplitude, $U^*$, is proportional to the sea surface gradient, $\varsigma_x$, divided by the sum of the inertial term $\omega U$ and the linearised friction term $FU$, where $\omega = 2\pi/P$, $P$ tidal period. Here 'single-point' analytical solutions for the associated vertical structure are derived for specified values of $\varsigma_x$. For the case of a constant eddy viscosity, $E$, this vertical structure is shown to be determined by two parameters $Y=D(\omega/E)^{1/2}$ and $J=3\pi (E\omega)^{1/2}/8 f U^*$ ($D$ water depth and $f$ bed friction coefficient). $Y$ is analogous to an Ekman height and $J$ introduces the effect of bed friction.

From both comparisons of observed structure against these solutions and self-consistency of the analytical solutions, the approximation $E=f U^* D$ is shown to be valid in strongly tidal, shallow waters where the influence of the bottom boundary layer extends to the surface. Figure 3.7 shows a comparison of tidal current structures determined using this constant value for $E$ versus detailed numerical simulations employing a turbulence closure module (Appendix 3B). The overall level of agreement indicates the validity of the approximation.

Adopting this approximation for $E$, the characteristics of vertical structure of tidal currents can be reduced to dependency on the Strouhal number, $S_R = UP/D$ with $Y \sim J \sim 50/S_R^{1/2}$ for $f=0.0025$. The amplitude structure becomes more pronounced with increasing $S_R$ up to an asymptotic limit of $S_R \sim 350$. Accompanying phase variations are a maximum for this value of $S_R$ but decrease for both smaller and larger values (Fig. 3.3). In meso- and macro-tidal estuaries, the Strouhal number will be well in excess of 1000.

Commonly observed features explained by this theory include, Figs. 3.1 and 3.3:

(1) depth-mean velocity occurring at fractional height ($z=Z/D$) above the bed, $z=0.4$;
(2) phase advance at the bed relative to the surface of up to 20° (at $S_R \sim 350$). A similar phase advance at the coastal boundaries relative to the deep mid-section generally occurs.
(3) maximum velocities occurring at the surface, except for $Y>4$ ($S_R < 300$) where sub-surface maxima occur (although not especially pronounced).

The above solutions ignore the effects of the Earth's rotation represented by the Coriolis term in (2.1) and (2.2). Even in long narrow estuaries, the Coriolis term is significant in determining the details of tidal current structure. This significance can be understood by noting that tidal currents do not simply ebb and flood along one axis but rotate in an elliptic pattern. The characteristics of these ellipses are represented by the parameters: $A_{MAX}$ amplitude along the principal axis, $A_{MIN}$ amplitude along the (orthogonal) minor axis, $\psi$ direction of the principal axis, $\varphi$ phase (time of maximum current). The eccentricity $E_C = A_{MIN}/A_{MAX}$ with the additional convention of positive for anti-clockwise (a–c) rotation and negative for clockwise (c–w).

Vertical variations in these ellipse parameters calculated from observations can often appear bewilderingly complex. However, an underlying systematic structure invariably emerges when the ellipse is separated into its clockwise and anti-clockwise rotational components. (1D flow corresponds to $E_C = A_{MIN} = 0$ and occurs when the magnitude of the two rotational components are equal). This separation enables the influence of the Coriolis term to be directly illustrated. Thus, expanding the above theory to include the Coriolis term introduces separate Strouhal Numbers for each rotational component as follows:

$$S_{a-c} = \frac{2\pi U}{D(\omega + \Omega)} \quad \text{and} \quad S_{c-w} = \frac{2\pi U}{D(\omega - \Omega)}. \tag{3.51}$$

For a semi-diurnal constituent, $\omega \sim 1.4 \times 10^{-4}$, while at a latitude of 50°, the Coriolis term $\Omega \sim 1.1 \times 10^{-4}$. Thus, for such conditions, we see that the c–w Strouhal number is generally an order of magnitude greater than the a–c value. This results in a much more pronounced vertical structure for the c–w component and directly explains the commonly observed features of

(1) increasing c–w eccentricity towards the surface
(2) major axis veering a–c towards the surface.

The acute sensitivity of tidal currents close to inertial latitudes, where $\omega = \Omega$ ($\sim$70° for semi-diurnal constituents and $\sim$30° for diurnal), is shown in Fig. 3.5. Although this may be qualified by evidence from Csanady (1976) that the thickness of the boundary layer and in consequence the vertical eddy viscosity is itself related to the Coriolis parameter.

The above mathematical approach used in determining the details of tidal current structure can be usefully applied to derive the steady-state vertical current structure associated with wind forcing (3.44). This theory explains the observed 'Ekman spiral' patterns of wind-driven currents shown in Fig. 3.8. Maximum surface currents, up to a few percent of the wind speed, occur in deep water, veering 45° to the right of the wind (northern hemisphere). In shallow water, bed friction reduces these currents and aligns them more closely with the wind direction.

## Appendix 3A

### 3A.1  Tidal current ellipse

Figure 3A.1 (Prandle, 1982b) illustrates how a current ellipse may be resolved into two rotary components. The current vectors are shown at time $t = 0$.

On the left, the anti-clockwise rotating component is given by

$$\boldsymbol{R}_1 = |\boldsymbol{R}_1|(\cos g_1 + \text{i} \sin g_1). \tag{3A.1}$$

In the middle, the clockwise rotating component is

$$\boldsymbol{R}_2 = |\boldsymbol{R}_2|(\cos g_2 + \text{i} \sin g_2). \tag{3A.2}$$

The maximum current occurs at time $t_{\text{MAX}}$ when $\boldsymbol{R}_1$ and $\boldsymbol{R}_2$ are aligned, i.e. when

74                              *Currents*

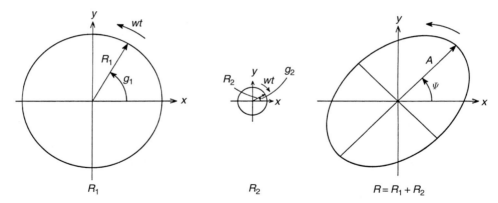

Fig. 3A.1. Decomposition of a tidal current ellipse into anti-clockwise and clockwise rotating components.

$$\omega t_{\text{MAX}} + g_1 = -\omega t_{\text{MAX}} + g_2$$

$$\text{or} \quad t_{\text{MAX}} = \frac{g_2 - g_1}{2\omega}. \tag{3A.3}$$

The phase, $\varphi$, is given by

$$\varphi = \omega t_{\text{MAX}} \tag{3A.4}$$

and the maximum amplitude, $A_{\text{MAX}}$, is

$$A_{\text{MAX}} = |R_1| + |R_2|. \tag{3A.5}$$

The inclination, $\psi$, of the maximum current to the $X$-axis is given by

$$\psi = \omega t_{\text{MAX}} + g_1 = -\omega t_{\text{MAX}} + g_2 = \frac{1}{2}(g_2 + g_1). \tag{3A.6}$$

The minimum amplitude, $A_{\text{MIN}}$, is when $R_1$ and $R_2$ are opposed, hence

$$A_{\text{MIN}} = ||R_1| - |R_2||. \tag{3A.7}$$

The eccentricity of the ellipse, $E_c$, is defined as the ratio $A_{\text{MIN}}$:$A_{\text{MAX}}$. However, in addition, we use the convention of $E_c$ positive for anti-clockwise current rotation $|R_1| > |R_2|$ and negative for clockwise rotation, $|R_2| > |R_1|$. Thus letting $\mu = |R_1|/|R_2|$, we define $E_c$ as follows:

$$E_c = \frac{\mu - 1}{\mu + 1}. \tag{3A.8}$$

Hence, $|E_c|$ is a minimum when $\mu = 1$ and $|E_c|$ increases as $\mu$ increases above this value.

## Appendix 3B

### *3B.1 Turbulence model*

The Mellor–Yamada level 2.5 mode (MYL 2 5, Mellor and Yamada (1974)) is widely used to determine the values of the vertical eddy viscosity, $K_M$, and vertical eddy diffusivity, $K_q$, coefficients. Refinements of this model described by Deleersnijder and Luyten (1993) are incorporated here. Umlauf and Burchard (2003) describe more recent developments.

The turbulent kinetic energy (TKE or $k$) equation is $k = q^2/2$

$$\frac{\partial}{\partial t}\left(\frac{q^2}{2}\right) = \frac{\partial}{\partial Z}\left[K_q \frac{\partial}{\partial Z}\left(\frac{q^2}{2}\right)\right] \quad \text{(diffusion)}$$

$$+K_M \left[\left(\frac{\partial U}{\partial Z}\right)^2 + \left(\frac{\partial V}{\partial Z}\right)^2\right] \quad \text{(shear production)}$$

$$-\frac{q^3}{B_1} = 0 \quad \text{(dissipation)}, \tag{3B.1}$$

where $B_1 = 16.6$ and the eddy coefficients are proportional to the mixing length $l$ and associated stability functions $S_q$ and $S_H$; thus

$$K_q = lq, \quad S_q = 0.2lq \tag{3B.2}$$

$$K_M = lq, \quad S_M = 0.4lq. \tag{3B.3}$$

The mixing length $l$ is determined from the related equation

$$\frac{\partial}{\partial t}q^2 l = \frac{\partial}{\partial Z}\left[K_q \frac{\partial}{\partial Z}\left(q^2 l\right)\right] + E_1 l\, K_M \left[\left(\frac{\partial U}{\partial Z}\right)^2 + \left(\frac{\partial V}{\partial Z}\right)^2\right] - \frac{Wq^3}{B_1}. \tag{3B.4}$$

The wall proximity function $W$ is defined as

$$W = 1 + \frac{E_2 l^2}{(K_v L)^2}, \tag{3B.5}$$

where $E_1 = 1.8$, $E_2 = 1.33$. $K_v$ is the van Karman constant ($\sim 0.4$), and $L$ is a function of the distance to the seabed $d_b$ and to the sea surface $d_s$; thus

$$L = \frac{d_s\, d_b}{d_s + d_b}. \tag{3B.6}$$

The boundary conditions specified at the surface and bed are

$$K_M \frac{\partial R}{\partial Z} = \frac{\tau_B}{\rho} \tag{3B.7}$$

$$K_M \frac{\partial R}{\partial Z} = \frac{\tau_0}{\rho}, \tag{3B.8}$$

where the velocity $R = U + iV$, $\tau_0$ is the applied wind stress and $\tau_B$ is the tidal stress at the bed $(\rho f R|R|)$.

At the surface and bed $q^2 l = 0$.

# References

Bowden, K.F., 1953. Note on wind drift in a channel in the presence of tidal currents. *Proceedings of the Royal Society of London*, A, **219**, 426–446.

Bowden, K.F., 1978. Physical problems of the Benthic Boundary Layer. *Geophysical Surveys*, **3**, 255–296.

Cheng, R.T., Feng, S., and Pangen, X., 1986. On Lagrangian residual ellipse. In: van de Kreeke, J. (ed.), *Physics of Shallow Estuaries and Bays* (Lecture Notes on Coastal and Estuarine Studies No. 16) Springer-Verlag, Berlin, pp. 102–113.

Csanady, G.T., 1976. Mean circulation in shallow seas. *Journal of Geophysical Research*, **81**, 5389–5399.

Davies, A.M. and Furnes, G.K., 1980. Observed and computed $M_2$ tidal currents in the North Sea. *Journal of Physical Oceanography*, **10** (2), 237–257.

Defant, A., 1961. *Physical Oceanography*, Vol. 1. Pergamon Press, London.

Deleersnijder, E. and Luyten, P., 1993. *On the Practical Advantages of the Quasi-Equilibrium Version of the Mellor and Yamada Level 2–5 Turbulence Closure Applied to Marine Modelling*. Contribution No. 69. Institui d'Astronomie et de Geophysique, Universite Catholique de Louvain, Belgium.

Faller, A.J. and Kaylor, R., 1969. Oscillatory and transitory Ekman boundary layers. *Deep-Sea Research, Supplement*, **16**, 45–58.

Flather, R.A., 1984. A numerical model investigation of the storm surge of 31 January and 1 February 1953 in the North Sea. *Quarterly Journal of the Royal Meteorological Society*, **110**, 591–612.

Geyer, W.R. and Signell, R., 1991. Measurements and modelling of the spatial structure of nonlinear tidal flow around a headland. In: Parker, B.B. (ed.), *Tidal Hydrodynamics*. John Wiley and Sons, New York, pp. 403–418.

Heaps, N.S., 1969. A two-dimensional numerical sea model. *Philosophical Transactions Royal Society, London*, A, **265**, 93–137.

Ianniello, J.P., 1977. Tidally-induced residual currents in estuaries of constant breadth and depth. *Journal of Marine Research*, **35** (4), 755–786.

Kraav, V.K., 1969. Computations of the semi-diurnal tide and turbulence parameters in the North Sea. *Oceanology*, **9**, 332–341.

Lane, A., Prandle, D., Harrison, A.J., Jones, P.D., and Jarvis, C.J., 1997. Measuring fluxes in estuaries: sensitivity to instrumentation and associated data analyses. *Estuarine, Coastal and Shelf Science*, **45** (4), 433–451.

Liu, W.C., Chen, W.B., Kuo, J-T., and Wu, C., 2008. Numerical determination of residence time and age in a partially mixed estuary using a three-dimensional hydrodynamic model. *Continental Shelf Research*, **28** (8), 1068–1088.

McDowell, D.M., 1966. Scale effect in hydraulic models with distorted vertical scale. *Golden Jubilee Symposia, Vol. 2, Central Water and Power Research Station, India*, pp. 15–20.

McDowell, D.M. and Prandle, D., 1972. Mathematical model of the River Hooghly. Proceedings of the American Society of Civil Engineers. *Journal of Waterways and Harbours Division*, **98**, 225–242.

Mellor, G.L. and Yamada, T., 1974. A hierarchy of turbulence closure models for planetary boundary layers. *Journal of the Atmospheric Science*, **31**, 1791–1806.

Munk, W., Snodgrass, F., and Wimbush, M., 1970. Tides off-shore: Transition from California coastal to deep-sea waters. *Geophysical Fluid Dynamics*, **1**, 161–235.

Pingree, R.D. and Maddock, L., 1980. Tidally induced residual flows around an island due to both frictional and rotational effects. *Geophysical Journal of the Royal Astronomical Society*, **63**, 533–546.

Prandle, D., 1975. Storm surges in the southern North Sea and River Thames. *Proceeding of the Royal Society of London*, A, **344**, 509–539.

Prandle, D., 1982a. The vertical structure of tidal currents and other oscillatory flows. *Continental Shelf Research*, **1**, 191–207.

Prandle, D., 1982b. The vertical structure of tidal currents. *Geophysical and Astrophysical Fluid Dynamics*, **22**, 29–49.

Prandle, D., 1984. A modelling study of the mixing of 137Cs in the seas of the European continental shelf. *Philosophical Transactions of the Royal Society I of London*, A, **310**, 407–436.

Prandle, D., 1991. A new view of near-shore dynamics based on observations from H.F. Radar. *Progress in Oceanography*, **27**, 403–438.

Prandle, D., 1997. The influence of bed friction and vertical eddy viscosity on tidal propagation. *Continental Shelf Research*, **17** (11), 1367–1374.

Prandle, D., 2004. Saline intrusion in partially mixed estuaries. *Estuarine, Coastal and Shelf Science*, **59**, 385–397.

Prandle, D. and Ryder, D.K., 1989. Comparison of observed (H.F. radar) and modelled near-shore velocities. *Continental Shelf Research*, **9**, 941–963.

Prandle, D. and Matthews, J., 1990. The dynamics of near-shore surface currents generated by tides, wind and horizontal density gradients. *Continental Shelf Research*, **10**, 665–681.

Prandle, D. and Player, R., 1993. Residual currents through the Dover Strait measured by H.F. Radar. *Estuarine, Coastal and Shelf Science*, **37** (6), 635–653.

Proudman, J., 1953. *Dynamical Oceanography*. Methuen and Co. Ltd, London.

Soulsby, R.L., Hamm, L., Klopman, G., Myrhaug, D., Simons, R.R., and Thomas, G.P., 1993. Wave–current interaction within and outside the bottom boundary layer. *Coastal Engineering*, **21**, 41–69.

Souza, A.J. and Simpson, J.H., 1996. The modification of tidal ellipses by stratification in the Rhine ROFI. *Continental Shelf Research*, **16**, 997–1008.

Umlauf, L. and Burchard, H., 2003. A generic length-scale equation for geophysical turbulence models. *Journal of Marine Research*, **6**, 235–265.

Van Veen J., 1938. Water movements in the Straits of Dover. *Journal du Conseil, Conseil International pour l'Exploration de la Mer*, **14**, 130–151.

Wolf, J. and Prandle, D., 1999. Some observations of wave–current interaction. *Coastal Engineering*, **37**, 471–485.

Zimmerman, J.T.F., 1978. Topographic generation of residual circulation by oscillatory (tidal) currents. *Geophysical and Astrophysical Fluid Dynamics*, **11**, 35–47.

# 4

# Saline intrusion

## 4.1 Introduction

The nature of saline intrusion in an estuary is governed by tidal amplitude, river flow and bathymetry. The pattern of intrusion may be altered by 'interventions' such as dredging, barrier construction or flow regulation alongside impacts from changes in msl or river flows linked to Global Climate Change. Adjustments to the intrusion may have important implications for factors such as water quality, sedimentation and dispersion of pollutants. Whereas tidal propagation can be explained from simple analytical formula and accurately modelled, it is often difficult to explain observed changes in intrusion from spring to neap or flood to drought.

The laterally averaged mass conservation equation may be written as (Oey, 1984)

$$\frac{\partial C}{\partial t} + U\frac{\partial C}{\partial X} + W\frac{\partial C}{\partial Z} = \frac{1}{DB}\left[\frac{\partial}{\partial X}\left(DBK_x\frac{\partial C}{\partial X}\right) + D\frac{\partial}{\partial Z}\left(BK_z\frac{\partial C}{\partial Z}\right)\right], \qquad (4.1)$$

where $C$ is concentration, $U$ and $W$ velocities in the axial $X$ and vertical $Z$ directions, $D$ water depth, $B$ breadth, $K_x$ and $K_z$ eddy dispersion coefficients. Lewis (1997) provides detailed descriptions of how advection and dispersion terms in (4.1) interact to promote estuarine mixing.

Dispersion of salt involves interacting 3D variations in phase, amplitude and mean values of both currents and the saline distribution. These variations are sensitive to the level of density stratification which may vary appreciably – temporally and both axially and transversally. The spectrum of such spatial and temporal variations includes

(1) the tidal cycle, with pronounced vertical mixing occurring on one or both of peak flood and ebb currents due to bottom friction, or at slack tides due to internal friction (Linden and Simpson, 1988);
(2) the neap–spring cycle, with mixing occurring more readily on spring than neap tides;

(3) the hydrological cycle, with variations in both river flow and salinity of sea water at the mouth (Godin, 1985);

(4) storm events including storm surges generated both internally and externally (Wang and Elliott, 1978) and surface wave mixing (Olson, 1986);

(5) variations in water density due to other parameters, in particular temperature (Appendix 4A) and suspended sediment load (Chapter 5).

This chapter does not consider the influence of temperature on density. An Appendix is included describing the seasonal cycle of temperature variations as a function of water depth and latitude. Nunes and Lennon (1986) indicate how evaporation at low latitudes can produce maximum salinities at the head of an estuary – inverting the customary pattern of saline intrusion.

### 4.1.1 Classification systems

Pritchard (1955) introduced a classification scheme for estuarine mixing. In class A, 'highly stratified estuaries', the dispersion terms on the right-hand side of (4.1) are negligible. For type B, 'partially mixed estuaries', vertical dispersion is important. Types C, wide, and D, narrow, are 'fully mixed estuaries' with density profiles $\partial C/\partial Z \sim 0$ and hence only longitudinal dispersion is involved. This latter assertion is inconsistent with subsequent notions of mixing in well-mixed estuaries, for example the role of tidal straining described in Section 4.4.

While Pritchard's classification system provides useful qualitative descriptions, in many estuaries the degree of stratification varies appreciably both spatially and temporally. Figure 4.1 shows a neap–spring time series of salinity variations at three depths for a single location in the Mersey Estuary. While the tidal signals are evident, longer-term (sub-tidal) influences are much greater than for corresponding elevation or current time series. Figure 4.2 (Liu *et al.*, 2008) shows how axial distributions in the Danshuei River vary with changing river flows.

The stratification diagram derived by Hansen and Rattray (1966) (Fig. 4.3) has been used to classify the nature of mixing in estuaries and the sensitivity of the stratification to changing conditions. The diagram is based on two parameters: (i) $\delta s/s$, the salinity difference between bed and surface divided by the depth-averaged salinity (mean values over a tidal cycle), and (ii) $U_s/\bar{U}$ residual velocity at the surface divided by the depth-mean value. Four estuarine types are identified with sub-divisions into (a) mixed and (b) stratified according to whether $\delta s/s$ is less or greater than 0.1. In estuaries of type 1, residual flow is seawards at all depths and consequently mixing of salt is entirely by diffusion. For type 2, residual flow reverses with depth and mixing is due to both advection and diffusion. For type 3, the vertical structure of the residual flow is so pronounced that advection accounts for over 99% of the mixing process; in type 3(b), mixing is confined to the

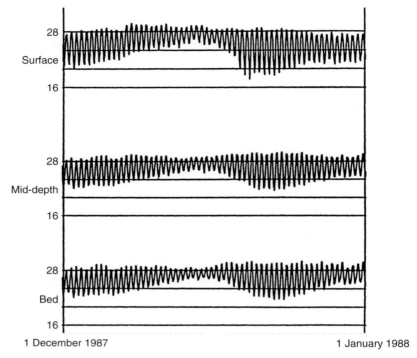

Fig. 4.1. Neap–Spring cycles of observed salinity variations, ‰, in the Mersey Narrows. Observed values at the surface, mid-depth and bed in water depth of 15 m.

near-surface region. Type 4 exhibits maximum stratification and approximates a salt wedge. Figure 4.11, derived from a 'single-point' numerical model of saline intrusion and mixing, emphasises the role of gravitational circulation in estuaries of type 4.

Prandle (1985) showed how the $U_s/\bar{U}$ axis could be replaced by $S/F$, i.e. the ratio of the residual accelerations associated with the horizontal density gradient and bed friction defined in (4.42), yielding a more direct assessment of the classification of an estuary based on more readily available parameters. The demarcation line which separates estuaries of types 1 and 2 can then be explained by the occurrence of flow reversal for $S/F > 24$ or $U_s/\bar{U} > 2$ (Table 4.1).

### 4.1.2 *Mixing in estuaries*

Dyer and New (1986) showed how the layer Richardson number, defined as

$$R_i = \frac{\dfrac{g}{\rho}\dfrac{\partial\rho}{\partial Z}}{\left(\dfrac{\partial U}{\partial Z}\right)^2}, \tag{4.2}$$

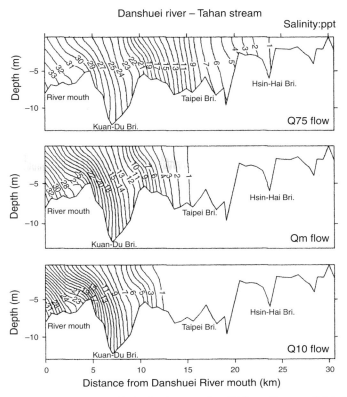

Fig. 4.2. Axial variations in salinity in the Danshuei River, Taiwan Q75, flow rate exceeded 75% of time, Q10 flow exceeded 10% of time.

representing the ratio of buoyancy forces to vertical turbulent force, determines the nature of mixing in estuaries. Vertical mixing occurs for $R_i < 0.25$ when turbulence is sufficient to overcome density layering. In determining whether an estuary is likely to be mixed or stratified, the question arises as to which components of current shear predominate (tidal, riverine or, circuitously, saline gradient-induced). Linden and Simpson (1988) and Simpson *et al.* (1990) have emphasised that stratification cannot be simply linked to the 'gross' parameters used in (4.2). The value of $R_i$ varies considerably along an estuary and over both the ebb–flood and the spring–neap cycles. We expect nascent stratification at certain times and locations in almost all estuaries. In Section 4.5, we examine conditions likely to sustain, and thereby consolidate, stratification showing how the ratio of riverine to tidal currents, $U_0/U^*$, is the clearest indicator of stratification.

Tidal advection can carry a fluid column several kilometres from its mean position. The total upstream and downstream excursion for a semi-diurnal tide is approximately $14U^*$ km, for tidal current amplitude, $U^*$, in m s$^{-1}$. Hence, stratification

Table 4.1 *Residual surface gradients and current components at the surface and bed*

|  | Surface gradient | Surface velocity | Bed velocity |
|---|---|---|---|
| (a) River flow $Q = U_0 D$ | $-0.89F$ | $1.14U_0$ | $0.70U_0$ |
| (b) Wind stress $\tau_W$ no net flow | $1.15W$ | $0.31\,(W/F)\,U_0$ | $-0.12\,(W/F)\,U_0$ |
| (c) Mixed density gradient | $-0.46S$ | $0.036\,(S/F)\,U_0$ | $-0.029\,(S/F)\,U_0$ |
| (d) Stratified 'wedge' | $-1.56F$ | $1.26U_0$ | $-0.18U_0$ |
| lower-layer depth $dH$ | $/(1-d)^2$ | $/(1-d)$ | $/(1-d)$ |

*Notes:* (a) $U_0$ river flow, (4.11) and (4.12), (b) $W$ wind forcing, (4.37) and (4.38), (c) $S$ mixed salinity, (4.15) and (4.16), (d) stratified salinity, (4,31), (4.32) and (4.35). $W$, $S$ and friction parameter, $F$, defined in (4.42).
*Source:* Prandle, 1985.

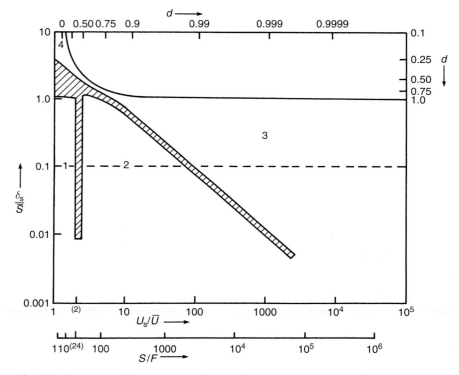

Fig. 4.3. Hansen and Rattray (1966) Stratification diagram, modified by Prandle (1985). $\delta s/s$ fractional salinity difference between bed and surface, $U_s/\bar{U}$ residual velocity at the surface: depth-averaged value, $S/F$ ratio of salinity to bed friction terms, $d$ fractional height of interface above the bed, (4.28) in a stratified system.

observed at one location represents an integration of processes acting over a wide area both axially and transversally. Abraham (1988) indicates that differential advection over depth $\partial/\partial Z \, (U(\partial \rho/\partial X))$ is important in determining stratification. This process is referred to as 'tidal straining', whereby larger currents at the surface on the flood tide carry denser sea water sufficiently far upstream that mixing occurs when the surface density exceeds that at the bed, is explored in Section 4.4. Conversely, on the ebb, the greater downstream movement of the surface waters enhances stability and promotes stratification. However, in highly stratified estuaries, vertical mixing may be a maximum on the ebb tide when current shear due to bottom friction and baroclinic forces augment mixing (Geyer and Smith, 1987).

In addition to tidal advection, observed stratification may be influenced by the propagation of internal waves generated elsewhere (New *et al.*, 1987). Abraham (1988) describes the differing mixing processes associated with external (tidal bed friction) and internal processes, the latter are shown to be important (around the times of slack water) for salt intrusion in the Rotterdam Waterway.

### *4.1.3 Present approach*

The present focus is on establishing analytical solutions to provide theoretical frameworks to explain and quantify, in scaling terms, the governing processes and thereby interpret detailed model studies. The emphasis is on mixed and partially mixed estuaries, this permits the assumption of a (temporally and vertically) constant relative axial density gradient, $S_x = (1/\rho)(\partial \rho/X)$, with density linearly proportional to salinity. The approach follows earlier solutions for tidally averaged linearised theories derived by Officer (1976), Bowden (1981) and Prandle (1985).

Section 4.2 examines the vertical structure of residual currents associated with saline intrusion (including the case of a saline 'wedge'), river flow and wind forcing. The vertical salinity profile, consistent with this current structure for saline intrusion, is also determined.

Section 4.3 examines both observational and theoretical approaches for predicting intrusion lengths. These include (i) flume studies, (ii) counter-balancing at the limit of intrusion, upstream residual velocity associated with $S_x$ with downstream river flow $U_0$ and (iii) a theoretical derivation for a stratified salt wedge. A further estimate of intrusion length is obtained from balancing the rate of mixing associated with vertical diffusion against river inflow. Most such studies apply to channels of constant breadth and depth. It is shown how the funnelling shape of estuaries introduces a further 'degree of freedom' in determining the axial location, and thereby the length, of the intrusion. This may account for widely reported difficulties in interpreting changes in intrusion over spring–neap and flood–drought cycles. The time lag involved in such axial shifts further complicates estimates of intrusion.

In Section 4.4, these tidally averaged linearised theories are extended to take account of tidal straining and associated convective overturning. Their applicability is evaluated by comparisons with a 'single-point' numerical model in which both the depth-averaged tidal current amplitude, $U^*$, and a (temporally and vertically) constant saline gradient, $S_x$, are specified. The model highlights the importance of convective overturning in counteracting unstable density structures introduced by tidal straining. To explore the generality of estuarine responses, the model is run for a range of values of saline intrusion lengths, $L_I$, and water depths, $D$.

Section 4.5 uses the above results to re-examine what indices best represent stratification levels in estuaries.

## 4.2  Current structure for river flow, mixed and stratified saline intrusion

Here we examine the residual current profiles associated with (i) river flow, (ii) a well-mixed longitudinal density gradient $\partial\rho/\partial Z$, (iii) a wedge-type intrusion with density difference $\Delta\rho$ and no vertical mixing at the interface and, for completeness, (iv) a surface wind stress $\tau_w$.

It is assumed that these separate motions may be described by linear equations, and hence, a complete flow description can be obtained by simple addition. For this assumption to be valid (i) the residual component of the bed friction term must be effectively linearised and (ii) the ratio of tidal elevation to mean water depth must be small. Condition (i) is satisfied in most tidal estuaries while condition (ii) tends merely to qualify the accuracy of the results in shallow, meso- and macro-tidal estuaries.

To provide useful quantitative results, two basic assumptions are made. First, we adopt a value for the bed stress coefficient $f = 0.0025$. Second, the eddy viscosity coefficient $E$ is assumed to be constant and given by (Prandle, 1982)

$$E = fU^*H, \tag{4.3}$$

where $U^*$ is the depth-mean tidal current amplitude and $H$ is total water depth.

Neglecting convective terms, the axial momentum equation at a point $Z$ above the bed is (3.6)

$$\frac{\partial U}{\partial t} + g\frac{\partial \varsigma}{\partial X} + g(\varsigma - Z)\frac{1}{\rho}\frac{\partial \rho}{\partial X} = \frac{\partial}{\partial Z}E\frac{\partial U}{\partial Z}, \tag{4.4}$$

where $\varsigma$ is the surface elevation and $\partial\rho/\partial X$ the density gradient, assumed here to be constant over both time and vertically.

### 4.2.1  Current structure for river flow Q

Omitting the density gradient, for steady axial flow (positive downstream), Equation (4.4) reduces to

$$g \, \frac{d\varsigma}{dX} = g \, i_Q = E \, \frac{\partial^2 U}{\partial Z^2},$$ 
(4.5)

where $i_Q$ is the residual elevation gradient associated with $Q$. By integrating (4.5) twice w.r.t. $Z$, the current profile is

$$U_Q = g \, \frac{i_Q}{E} \, \left( \frac{Z^2}{2} - HZ - \frac{EH}{\beta} \right),$$ 
(4.6)

where the constants of integration were determined from the two boundary conditions:

$$\text{stress at the surface} \quad \tau_{Z=H} = \rho E \frac{\partial U}{\partial Z} = 0$$ 
(4.7)

$$\text{stress at the bed} \quad \tau_{Z=0} = \rho E \frac{\partial U}{\partial Z} = \rho \beta U_{Z=0}.$$ 
(4.8)

Condition (4.8) applies when $U^* \gg U_Q$, in which case Bowden (1953) showed that $\beta = \left( \frac{4}{\pi} \right) f \, U^*$.

Finally, introducing the depth-averaged velocity $\bar{U}_Q = \frac{Q}{H} = \frac{1}{H} \int_0^H U \, dZ$, Equation (4.6) gives

$$i_Q = \frac{\bar{U}_Q}{g \left( \dfrac{H^2}{3E} + \dfrac{H}{\beta} \right)}$$ 
(4.9)

and

$$U_Q = \bar{U}_Q \, \frac{\left\{ -\dfrac{z^2}{2} + z + \dfrac{E}{\beta H} \right\}}{\dfrac{1}{3} + \dfrac{E}{\beta H}}$$ 
(4.10)

with $z = Z/H$

By inserting (4.3), (4.9) and (4.10) reduce to

$$i_Q = -0.89 \, \frac{f \, \bar{U} \, U^*}{g \, H}$$ 
(4.11)

and

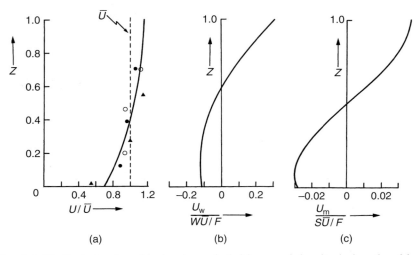

Fig. 4.4. Vertical structure for riverine, wind-driven and density-induced residual currents. (a) Freshwater flow $Q$ (4.12), point values indicate observed data from three positions; (b) wind stress $\tau_w$ (4.37); (c) well-mixed longitudinal density gradient (4.15). $W$, $F$ and $S$ defined in (4.42).

$$U_Q = 0.89\,\bar{U}_Q\left(\frac{-z^2}{2} + z + \frac{\pi}{4}\right). \tag{4.12}$$

This residual current profile for river flow is illustrated in Fig. 4.4(a) (Prandle, 1985).

### 4.2.2 *Current structure for a well-mixed horizontal density gradient*

Again omitting the inertial term before adding a well-mixed longitudinal density gradient $\partial\rho/\partial x$, the steady state form of (4.4) yields

$$U = g\frac{d\varsigma}{dX}\frac{H^2}{E}\left(\frac{z^2}{2} - z - \frac{E}{H\beta}\right) + \frac{g}{\rho}\frac{\partial\rho}{\partial X}\frac{H^3}{E}\left(-\frac{z^3}{6} + \frac{z^2}{2} - \frac{z}{2} - \frac{E}{2H\beta}\right). \tag{4.13}$$

To isolate the influence of the density gradient, subscript M, we define

$$\frac{d\varsigma}{dX} = i_Q + i_M. \tag{4.14}$$

Then subtracting $U_Q$ given by (4.10) from (4.13) and applying the condition $\bar{U}_M = 0$ and the eddy viscosity formulation (4.3) yields the residual velocity structure for 'mixed' intrusions:

$$U_M = \frac{g}{\rho}\frac{\partial\rho}{\partial X f}\frac{H^2}{U\,U^*}\left(-\frac{z^3}{6} + 0.269\,z^2 - 0.037\,z - 0.029\right) \tag{4.15}$$

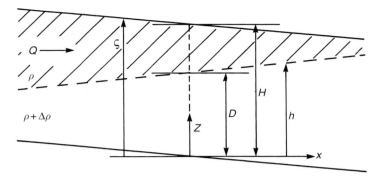

Fig. 4.5. Notation for a stratified saline wedge.

$$\text{and} \qquad i_M = -0.46 \, \frac{H}{\rho} \frac{\partial \rho}{\partial X}. \tag{4.16}$$

The current profile (4.15) is illustrated in Fig. 4.4(c). The associated net increase in mean sea level over the intrusion length is then approximately $0.013H$.

### 4.2.3 Current structure for a stratified two-layer density regime

Using the notation shown in Fig. 4.5, the pressure at height $Z$ is

in the layer $Z \geq D$, $\qquad p = \rho g(\varsigma - Z)$ $\tag{4.17}$

in the bottom layer $Z < D$, $p = \rho g(\varsigma - D) + (\rho + \Delta\rho)g\,(D - Z)$. $\tag{4.18}$

In addition to the earlier boundary conditions (4.7) and (4.8), we require continuity of both velocity and stress at the interface (subscript I); thus at $Z = D$,

$$U_I = U_B = U_T \tag{4.19}$$

$$\text{and} \qquad \rho \, E_T \frac{\partial U_T}{\partial Z} = (\rho + \Delta\rho) \, E_B \frac{\partial U_B}{\partial Z}, \tag{4.20}$$

where subscripts T and B denote values in the top and bottom layers, respectively.

Further assuming:

(a) $\rho + \Delta\rho \sim \rho$ in mass calculations not involving buoyancy effects,
(b) zero net flow in the lower layer, and net flow $Q$ in the top layer,
(c) eddy viscosity in the lower layer given by a modified form of (4.3), namely,

$$E_B = f U^* dH \tag{4.21}$$

with $d = D/H$,
(d) eddy viscosity in the top layer given by

$$E_T = \gamma E_B. \tag{4.22}$$

Then proceeding as in earlier sections, we obtain

$$U_T = -\frac{Q\varepsilon}{Hd}\left[\frac{1}{\gamma}\left(\frac{z^2}{2} - z + d - \frac{d^2}{2}\right) - 0.308d(1-d)\right] \tag{4.23}$$

$$U_B = -\frac{Q\varepsilon}{H}\frac{(1-d)}{d^2}\left(-0.574\,z^2 + 0.149\,zd + 0.117d^2\right) \tag{4.24}$$

$$i_s = \frac{d\varsigma}{dX} = -\frac{Q\varepsilon f U^*}{gH^2}, \tag{4.25}$$

where

$$\varepsilon = \frac{d}{(1-d)^2\left[\frac{1}{3\gamma}(1-d) + 0.308d\right]} \tag{4.26}$$

$$a = 0.149 - 1.149d^{-1} \tag{4.27}$$

with

$$a = \frac{\frac{\Delta\rho}{\rho}\frac{dh}{dX}}{\frac{d\varsigma}{dX}}. \tag{4.28}$$

From expressions (4.27) and (4.28), the slope of the interface, $dh/dX$, is always greater and opposite in sign to the product of surface slope and density difference, $(\Delta\rho/\rho\, d\varsigma/dX)$ except at the surface, $d=1$, where they are of equal magnitude.

It is difficult to determine a general formulation for $\gamma$. Prandle (1985) examined various possibilities against observed current profiles and suggested the following expedient:

$$\gamma = \frac{1-d}{d}. \tag{4.29}$$

This permits the additional simplifications:

$$E_T = f\,U^*(H - D) \tag{4.30}$$

$$i_S = \frac{-1.56}{(1-d)^2}\frac{Q f U^*}{g\,H^2} \tag{4.31}$$

$$U_T = -1.56\frac{Q}{H}\frac{1}{(1-d)^3}\left(\frac{z^2}{2} - z - 0.808d^2 + 1.616d - 0.308\right) \tag{4.32}$$

$$U_B = -1.56\frac{Q}{H}\frac{1}{d^2(1-d)}\left(-0.574\,z^2 + 0.149\,zd + 0.117\,d^2\right). \tag{4.33}$$

### 4.2.4 Current structure for a constant surface wind stress $\tau_w$

Proceeding as for net flow $Q$, but with $\bar{U}_W = 0$ and the surface boundary condition

$$\tau_W = \rho E \frac{\partial U}{\partial Z},\qquad(4.34)$$

where $\tau_w$ is an imposed wind stress and subscript W is used to denote wind-driven components, we obtain

$$U_W = \frac{\tau_W H}{\rho E}\cdot\frac{\left[\frac{z^2}{2}\left(\frac{1}{2}+\frac{E}{\beta H}\right)-\frac{z}{6}-\frac{1}{6}\frac{E}{\beta H}\right]}{\left(\frac{1}{3}+\frac{E}{\beta H}\right)}\qquad(4.35)$$

$$116pt\text{and}\qquad i_W = \frac{\tau_W}{\rho g H}\frac{\left(0.5+\dfrac{E}{\beta H}\right)}{\dfrac{1}{3}+\dfrac{E}{\beta H}}\qquad(4.36)$$

or by introducing (4.3)

$$U_W = \frac{\tau_W}{\rho f U^*}\left(0.574\,z^2 - 0.149\,z - 0.117\right)\qquad(4.37)$$

and

$$i_W = \frac{1.15\tau_w}{\rho g H}.\qquad(4.38)$$

The factor 1.15 in (4.38) lies between the value 1.5 obtained for a no-slip bed condition and 1.0 for a full-slip condition by Rossiter (1954). The current profile (4.37) is illustrated in Fig. 4.4(b).

### 4.2.5 Time-averaged vertical salinity structure (mixed)

Scaling analysis can be used to justify the neglect of $W\,\partial C/\partial Z$ in (4.1), and observations indicate the predominance of the vertical diffusion term. Thence, assuming $\partial C/\partial X = S_x$, omitting time varying terms, we obtain, for $K_z$ constant, the salinity structure:

$$s(Z) = \int\int \rho\,U_M\,\frac{S_x}{K_z}\,dZ\,dZ,\qquad(4.39)$$

i.e. from (4.15), defining vertical variations in salinity as $s'(Z) = s(Z) - \bar{S}$,

$$s' = \rho\frac{g\,S_x^2}{E_z K_z}\frac{D^5}{10\,000}\left(-83\,z^5 + 224\,z^4 - 62\,z^3 - 146\,z^2 + 33\right).\qquad(4.40)$$

This time-averaged salinity profile is shown in fig. 4.8.

### 4.2.6 Summary

Figure 4.4 shows the residual current profiles (4.12), (4.15) and (4.37) pertaining to river flow, mixed saline intrusion and wind forcing. Table 4.1 summarises these results showing corresponding values at the surface and bed along with related gradients in surface elevation.

To interpret the magnitudes of these residual flow components, we introduce the depth-averaged equation of motion for steady state residual flow:

$$\frac{\partial \zeta}{\partial X} + \frac{H}{2\rho}\frac{\partial \rho}{\partial X} - \frac{\tau_w}{\rho g H} + \frac{4}{\pi}\frac{fU^*U_0}{gH} = 0. \tag{4.41}$$

Then, introducing the dimensionless parameters

$$S = H\frac{\partial \rho}{\rho \partial X}, \qquad W = \frac{\tau_w}{\rho g H}, \qquad F = \frac{f\,U^*U_0}{gH} \tag{4.42}$$

from (4.41), the forcing terms associated with density, wind and bed friction are in the ratio

$$\frac{S}{2} \; : \; W \; : \; \frac{4}{\pi}F. \tag{4.43}$$

From Table 4.1, the density forcing term $S/2$ is balanced by a surface gradient $-0.46S$ with the remaining component driving a residual circulation of $0.036\,SU_0/F$ seawards at the surface and $-0.029\,SU_0/F$ (landwards) at the bed. Similarly, the wind stress term $W$ is counteracted by a surface gradient $1.15W$ with the 'excess' balance driving a circulation of $0.31\,WU_0/F$ at the surface and $-0.12\,WU_0/F$ at the bed. Clearly, wind forcing is more effective in producing a residual circulation than longitudinal density gradients and both forcings influence elevations to a greater extent than currents.

## 4.3 The length of saline intrusion

### 4.3.1 Experimental derivation

Rigter (1973) carried out an extensive study of intrusion lengths both in a laboratory flume and in the Rotterdam Waterway. The flume was of constant depth and breadth. Analysis of the experiments resulted in an expression for saline intrusion length (Prandle, 1985) as follows:

$$L_I = \frac{0.18\,g(\Delta\rho/\rho)D^2}{f'\,U^*\,U_0} = \frac{0.005\,D^2}{f\,U^*\,U_0}, \tag{4.44}$$

where the latter applies for real estuaries with $\Delta\rho/\rho \sim 0.027$. Prandle (2004b) indicates that to maintain turbulent Reynolds numbers in flume studies, the friction

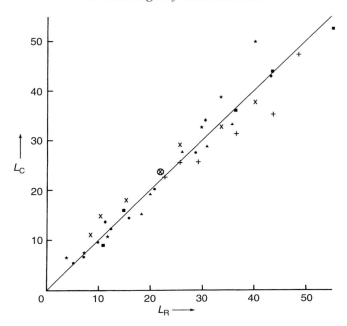

Fig. 4.6. Computed versus observed salinity intrusion lengths. Computed, $L_C$ (4.44); observed, $L_R$, from flume tests; $\otimes$ reference test, varying: $\times$ tidal amplitude, $\blacklozenge$ density difference, $\blacktriangle$ bed friction, $\bigstar$ river flow, $\blacksquare$ water depth, $+$ channel length.

factor must be increased by the same amount as the vertical scale exaggeration, i.e. $f' = 10f$.

A comparison of observations from this flume study and computed values for $L_I$ from (4.44) is shown in Fig. 4.6 (Prandle, 1985) for a range of tidal amplitudes, density differences, bed friction coefficients, river flows, water depths and flume lengths. The excellent agreement is confirmed by a correlation coefficient calculated as $R = 0.97$, indicating the robustness of (4.44) over a range of parameter sensitivity tests.

### 4.3.2 Derivations from velocity components

A simple hypothesis for the limit of upstream intrusion is the position where a balance exists at the bed between (i) the upstream velocity associated with a well-mixed salinity gradient and (ii) the downstream river velocity. From Table 4.1, this requires

$$0.029 \frac{g \, D^2 \, S_x}{f \, U^*} = 0.7 \, U_0 \tag{4.45}$$

i.e. $\qquad L_I = \dfrac{0.011 \, D^2}{f \, U^* \, U_0}, \tag{4.46}$

where $S_x = 0.027/L_I$.

### 4.3.3  Derivation for a stratified 'salt wedge'

From integration of the slope of the interface, Prandle (1985) derived the following expression for the length of intrusion of a stratified saline wedge

$$L_I = 0.26 \, \frac{g \, D^2}{f \, U^* \, U_0} \, \frac{\Delta\rho}{\rho} = \frac{0.07 \, D^2}{f \, U^* \, U_0}, \tag{4.47}$$

where $\Delta\rho/\rho \sim 0.027$.

This result may be compared with the following expression for the length, $L_A$, of an arrested saline wedge given by Keulegan (Ippen, 1966, Ch 1):

$$L_A = A \, \frac{g^{5/4} \, D^{9/4}}{U_0^{5/2}} \, \left(\frac{\Delta\rho}{\rho}\right)^{5/4}, \tag{4.48}$$

where $A$ is a parameter which varies with river conditions. Since Keulegan's expression was derived from observational data, the correspondence with (4.47) adds useful support to the present theoretical result.

Officer (1976, Ch 4) provides alternative derivations for the length of an arrested salt wedge, the results obtained obviously depend on the particular dynamical assumptions.

### 4.3.4  Mean value for saline intrusion length, $\mathbf{L}_I$

The above formulations (4.44), (4.46) and (4.47) all show identical expressions for $L_I$ but with quantitative values varying by the respective coefficients 0.005, 0.011 and 0.07. To reconcile these, we note that the balancing of sea-bed velocity components in (4.46) should strictly use values of $D$ and $U$ at the landward limit of the wedge, resulting in a decrease in the coefficient shown. Likewise, it is widely observed that intrusion lengths in stratified conditions are significantly longer than in mixed. Thus, henceforth we adopt the value (4.44).

A subsequent derivation (4.59), based on balancing the mixing rates associated with the density structure (4.54) with river flow (4.57), provides a further estimate for $L_I$ in close agreement with (4.44).

#### Observed versus computed values

A major difficulty in assessing the validity of (4.44) is the paucity of accurate observational data. Practical application of these estimates for saline intrusions lengths are complicated by a wide spectrum of associated response times. These range from minutes for turbulent intensity levels to hours for effective vertical mixing (Section 4.5) and days for estuarine flushing (4.60). Here, we utilise the observational data provided by Prandle (1981) from six estuaries (eight data sets) summarised in

Table 4.2 *Estuarine parameters*

|  | $n$ | $m$ | $L$(km) | $L_I$(km) | $D$(m) | $B$(km) | $Q$(m$^3$ s$^1$) | $\varsigma$*(m) |
|---|---|---|---|---|---|---|---|---|
| (A) Hudson | 0.7 | 0.4 | 248 | 99 | 11.6 | 3.7 | 99 | 0.8 |
| (B) Potomac | 1.0 | 0.4 | 184 | 74 | 8.4 | 18 | 112 | 0.7 |
| (C) Delaware | 2.2 | 0.3 | 214 | 43 | 4.4 | 28 | 300 | 0.6 |
| (D) Bristol Ch. | 1.7 | 1.2 | 138 | 55 | 29.3 | 20 | 80 | 4.0 |
| (E) Bristol Ch. |  |  |  | 138 |  |  | 8500 | 4.0 |
| (F) Thames | 2.3 | 0.7 | 95 | 76 | 12.6 | 7 | 480 | 2.0 |
| (G) Thames |  |  |  | 38 |  |  | 19 | 2.0 |
| (H) St. Lawrence | 1.5 | 1.9 | 48 | 167 | 74 | 48 | 210 | 1.5 |

*Notes:* $L$ estuarine length, $L_I$ observed intrusion length; $n$, $m$ breadth and depth variations ($x^n$, $x^m$), $Q$ river flow, $D$ depth, $B$ breadth and $\varsigma$* tidal amplitude at the mouth.
*Source:* Prandle, 1981.

Table 4.2. This data set provides estimates for intrusion length, $L_I$, riverflow, $Q$, and estuarine bathymetry. Values for $D$ and $U_0$ at the centre of the intrusion were estimated from power series approximations to breadth ($x^n$) and depth ($x^m$). The values for $U$* are estimated from (6.9) derived for synchronous estuaries.

Figure 4.7 (Prandle, 2004b) shows estimates of $L_I$ from (4.44) based on values for $U_0$, $U$* and $D$ at successive positions along the estuaries listed in Table 4.2. These values are summarised in Table 4.3, indicating values of $U_0$, $U$* and $D$ at the locations $X_c$, where the value of $L_I$ from (4.44) equals the observed value $L_0$. It is encouraging to note (Table 4.3) that the values of $X_c$ are in reasonable agreement with the related observational values $X_0$. Values of $U_0$ range from 0.17 to 0.57 cm$^{-1}$ with the exception of the St. Lawrence where $U_0 = 1.4$ cm$^{-1}$.

### 4.3.5 *Axial location of saline intrusion*

Figure 4.7 also indicates the landward limits of saline intrusion $X_u = (X_c - L_I/2)/L$ corresponding to successive values of $X_c$. We note that, with the exception of the Hudson and Delaware, the locations of $X_c$, where observed and computed values of $L_I$ are equal, correspond with or are slightly seawards of maximum values of $X_u$, i.e. where the landward limit of saline intrusion is a minimum.

Adopting this latter result as a criterion to determine the position, $x_i$, where the saline intrusion will be centred, requires in dimensionless terms, with $x = X/L$:

$$\frac{\partial}{\partial x}(x - 0.5 l_i) = 0, \tag{4.49}$$

where $l_i = L_I/L$. Utilising (4.44) and introducing the shallow water approximation, (6.9), for current amplitude in relation to elevation amplitude, $\varsigma$*,

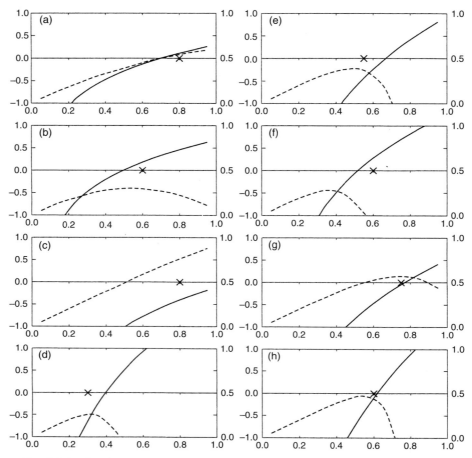

Fig. 4.7. Ratio of computed, $L_c$, to observed, $L_0$, saline intrusion lengths at varying locations, $X_c$. Results for (a) Hudson (b) Potomac (c) Delaware (d) Bristol Channel, $Q = 80\,\text{m}^3\,\text{s}^{-1}$ (e) Bristol Channel, $Q = 480\,\text{m}^3\,\text{s}^{-1}$ (f) Thames, $Q = 19\,\text{m}^3\,\text{s}^{-1}$ (g) Thames, $Q = 210\,\text{m}^3\,\text{s}^{-1}$ and (h) St. Lawrence. Horizontal axis, $X_C/L$. Point value X is the centre of observed intrusion. Vertical axis, (LHS) $\log_{10} L_C/L_0$, full line, (RHS) landward limit of intrusion; $X_U = (X_C - L_I/2)/L$, dashed line; $L_C$ from (4.44).

$$U^{*2} = \frac{\varsigma^* \omega (2gD)^{1/2}}{1.33 f}. \tag{4.50}$$

Then further assuming $Q = U_0\, D_i^2 / \tan\alpha$, where $\tan\alpha$ is the side slope for a triangular cross section, we obtain

$$x_i^2 = \frac{333\, Q \tan\alpha}{D_0^{5/2}}. \tag{4.51}$$

Table 4.3 *Observed and computed estuarine parameters for saline intrusion*

|  | $X_0$ | $X_c$ | $U_0$ (cm s$^{-1}$) | $U^*$ (m s$^{-1}$) | $D$ (m) |
|---|---|---|---|---|---|
| (A) Hudson | 0.80 | 0.70 | 0.35 | 0.59 | 9.7 |
| (B) Potomac | 0.60 | 0.50 | 0.20 | 0.54 | 6.3 |
| (C) Delaware | 0.80 | 1.0 | 0.29 | 0.47 | 4.4 |
| (D) Bristol Ch. | 0.30 | 0.40 | 0.22 | 1.53 | 9.5 |
| (E) Bristol Ch. | 0.55 | 0.65 | 0.28 | 1.70 | 17.3 |
| (F) Thames | 0.60 | 0.55 | 0.17 | 1.01 | 8.0 |
| (G) Thames | 0.75 | 0.50 | 0.57 | 1.06 | 10.2 |
| (H) St. Lawrence | 0.60 | 0.65 | 1.4 | 0.80 | 30.0 |

*Notes: $X_0$* (fraction of *L*) centre of observed intrusion, $X_c$ centre of intrusion when $L_I$ (4.44)$= L_0$, observed; *D*, $U_0 = Q$/area, $U^*$ depth, residual and tidal currents at $X_c$.
*Source:* Prandle, 2004a.

By further introducing the derivation for a synchronous estuary in Chapter 6 that depths and breadths vary with $x^{0.8}$, the depth, $D_i$ at $x_i$ is $D_0 x_i^{0.8}$ giving a residual velocity $U_0$ at the centre of the intrusion:

$$U_0 = \frac{D_i^{1/2}}{333} \text{ m s}^{-1}. \tag{4.52}$$

This expression yields values of $U_0 = 0.006$ m s$^{-1}$ for $D = 4$ m, 0.012 m s$^{-1}$ for $D = 16$ m.

Noting that (4.51) corresponds to $l_i = 2/3 x_i$, these values for $U_0$ will increase by a factor of 2 at the upstream limit and decrease by 40% at the downstream limit. Noting also the inaccuracy inherent in measurements of $U_0$, we conclude that these estimates of residual velocity associated with river flow in the saline intrusion region are reasonably consistent with the observed values shown in Table 4.3.

If in proceeding from (4.49) to (4.51) we introduce estuarine bathymetry with breadth $B_0 x^n$ and depth $D_0 x^m$, we obtain the following alternative form for (4.51)

$$x_i = \left( \frac{855 \, Q}{D_0^{3/2} B_0 \, (11m/4 + n - 1)} \right)^{1/(11m/4)+n-1}. \tag{4.53}$$

An especially interesting feature of the results for the axial location of saline intrusion (4.51) and (4.53) and the expression for residual river flow current (4.52) is their independence of both tidal amplitude and bed friction coefficient. (Although there is an implicit requirement that tidal amplitude is sufficient to maintain partially mixed conditions.) Equations (4.51) and (4.53) emphasise how the centre of the intrusion adjusts for changes in river flow $Q$. This 'axial migration' can severely complicate the sensitivity of saline intrusion beyond the anticipated direct responses apparent from the expression (4.44) for the length of intrusion, $L_I$.

## 4.4 Tidal straining and convective overturning

The preceding studies, involving tidally averaged linearised theories relating to the vertical structure of salinity and velocities, are now extended by numerical simulations incorporating tidal straining and associated convective overturning.

### 4.4.1 Rates of mixing

Before describing the modelling study, we first examine rates of mixing associated with (i) time-averaged density structure, (ii) tidal straining and (iii) supply of freshwater velocity.

From (4.1), the mean rate of mixing $M_K$ associated with the time-averaged density structure (4.40) is (Simpson *et al.*, 1990)

$$M_K = \rho \int_0^D \left| \frac{\partial}{\partial Z} K_z \frac{\partial s}{\partial Z} \right| \, dZ = 0.02 \rho g \frac{S_x^2 D^4}{E}. \tag{4.54}$$

The mean rate of overturning to compensate for tidal straining on the flood phase $M_0$ is

$$M_0 = \rho \frac{2}{\pi} U^* S_x D \int_0^1 |(0.7 + 0.9z - 0.45z^2 - 1) \, dz \tag{4.55}$$

$$= 0.12 \rho S_x D U^*, \tag{4.56}$$

where the approximation for the vertical structure of tidal currents is similar to that of Bowden and Fairbairn (1952). Nunes Vaz and Simpson (1994) derive additional contributions to vertical mixing associated with wind, surface heat exchange and evapo-transpiration.

For a stationary salinity distribution, the mixing rate $M_Q$ to balance freshwater velocity $U_0$ is

$$M_Q = \rho U_0 S_x D. \tag{4.57}$$

Then to balance $M_Q$, we expect values of $U_0$ of

tidal straining on flood tide $\qquad\qquad 0.12 U^* \qquad$ (4.58)

vertical salinity difference $\qquad\qquad 0.02 g \dfrac{S_x D^3}{E}. \qquad$ (4.59)

For $U^* = 0.5\,\mathrm{m\,s}^{-1}$, assuming no mixing on the ebb tide, (4.58) indicates a value of $U_0 = 3\,\mathrm{cm\,s}^{-1}$. Since this value generally exceeds that associated with vertical salinity differences (4.59), we expect that tidal straining will often eliminate vertical salinity differences by the end of the flood tide, see Fig. 4.10. Equation (4.59) provides an expression for saline intrusion length $L_I$ almost identical to (4.44).

For a prismatic channel, the estuarine flushing time, $T_F$ can be approximated by the time taken to replace half of the fresh water in the intrusion zone by river flow, i.e.

$$T_F = \frac{0.5(L_I/2)}{U_0}.\qquad(4.60)$$

For typical values of $T_F = 2$–$10$ days, this requires $U_0$ varying from 1.5 to 3 cm s$^{-1}$ over the range of intrusion lengths $L_I = 12.5$ to 100 km considered in Figs 4.11 and 4.12.

### 4.4.2 Modelling approach (Prandle, 2004a)

The significance of tidal straining is evaluated by reference to a 'single-point' numerical model in which the time-varying cycle of depth-averaged tidal current amplitude, $U^*$, and a (temporally and vertically) constant saline gradient, $S_x$, are specified (Prandle, 2004a). The model illustrates the role of convective overturning in counteracting unstable density structures introduced by tidal straining. The validity of the model is evaluated by simulation of measurements by Rippeth *et al.* (2001).

To explore the generality of estuarine responses, the model is run for a wide range of values of saline intrusion lengths, $L_I$, and water depths, $D$. Additional sensitivity analyses are made for changes in $U^*$ and the bed stress coefficient, $f$. The model is used to calculate the following parameters: surface-to-bed differences in both residual velocity, $\delta u$, and salinity, $\delta s$; the potential energy anomaly $\phi_M$ (4.65); the ratio of mixing by diffusivity: overturning; the efficiency of mixing $\varepsilon$ and the river flow $U_0$ (required to balance mixing rates). Thus, the model simulations illustrate, over a wide range of estuarine conditions, the impact of tidal straining and, thereby, the significance of its omission in the analytical formulations.

### 4.4.3 Model formulation

We assume (vertically and temporally) constant coefficients for eddy viscosity and diffusivity; however, a partial assessment of buoyancy modulation of eddy viscosity and diffusivity is examined by introducing the formulations of Munk and Anderson (1948):

$$\text{eddy viscosity} \qquad E_z = f\,U^*\,D\,(1 + 10R_i)^{-1/2} \qquad (4.61)$$

$$\text{eddy diffusivity} \qquad K_z = f\,U^*\,D/S_c\,(1 + 3.33R_i)^{-3/2}, \qquad (4.62)$$

where the Richardson number, $R_i$, is given by (4.2) and $S_c$ is the Schmidt number representing the ratio $E : K_z$. Nunes Vaz and Simpson (1994) describe a wide range of such formulations.

A single-point numerical model was formulated providing solutions to (4.1) and (4.4). The model is forced via specification of the depth-averaged semi-diurnal tidal current amplitude, $U^*$, and the (temporally and vertically constant) salinity gradient, $S_x$. The model was run for a (small) number of tidal cycles to achieve cyclic convergence.

Boundary conditions were specified as zero stress at the surface and bed stress

$$\tau_Z = \rho\,f\,U^2_{z=0}. \qquad (4.63)$$

The model incorporates convective overturning whenever density decreases with depth.

To distinguish the effect of (i) buoyancy modulation of $E$ and $K_z$ (via (4.61) and (4.62)) and (ii) convective overturning, three separate simulations were made. The first simulation used constant values of $E = fU^*D$ and $K_z = S_cE$. The second modified these values according to (4.61) and (4.62) and the third introduced convective overturning (in addition to (4.61) and (4.62)). Results from the first two simulations showed little difference, except for applications involving pronounced stratification where (4.1) and (4.4) become invalid. Hence, results are only shown for (i) Simulation 1 where $E = fU^*D$, $K_z = S_cE$ without convective overturning and (ii) Simulation 3 with $E$ and $K_z$ from (4.61) and (4.62) and convective overturning included.

### 4.4.4 Model results

#### Currents

Figure 4.8(a) shows tidally averaged (residual) current profiles obtained from Simulations 1 and 3 together with the theoretical profile (4.15). The parameters, $U^* = 0.6\,\text{m s}^{-1}$ $f = 0.0025$, $D = 32\,\text{m}$, $L = 400\,\text{km}$, $S_c = 0.1$, were used to correspond to the observations by Rippeth *et al.* (2001). Without overturning, the model accurately reproduced the earlier theoretical vertical profiles of both tidal current amplitude and phase. The results for model Simulation 1 and the theoretical solution (4.15) are near-identical. By contrast, the results for Simulation 3 show similar profiles but with the magnitude doubled.

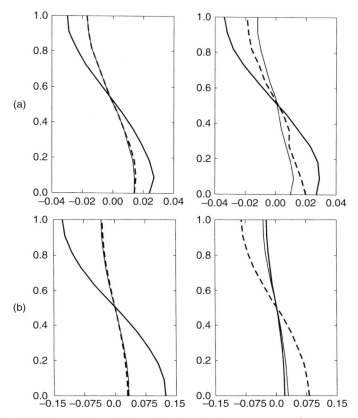

Fig. 4.8. Vertical profiles of (a) residual velocity (4.15), m s$^{-1}$, and (b) salinity differences (4.40), ‰. Left: dashed lines, theory; full lines, numerical model; thin, Simulation 1; thick, Simulation 3. Right: Simulation 3 with thin, $f = 0.01$; dashed, $U^* = 1.2$ m s$^{-1}$; thick $S_c = 1$. Values correspond to $U^* = 0.6$ m s$^{-1}$, $f = 0.0025$, $D = 32$ m, $L = 400$ km, $S_c = 10$.

Sensitivity analyses of Simulation 3 showed little variation with changes in the Schmid number. However, doubling $U^*$ produces a one-third reduction in residual velocities while quadrupling $f$ produces a two-third reduction.

### *Salinity*

Figure 4.8 (Prandle, 2004a) shows tidally averaged salinity profiles (converted to psu) for these same simulations compared with the theoretical profile from (4.40). Again, the theoretical result is near-identical to Simulation 1. However, Simulation 3 shows results four times larger with a net surface-to-bed value of $\delta s = 0.25$‰ similar to the observed values indicated by Rippeth *et al.* (2001).

Figure 4.9 (Prandle, 2004a) shows a tidal cycle of dissipation due to (i) bottom friction $\rho f U^3 / D$ and (ii) depth-averaged vertical shear $\rho E(\partial U / \partial Z)^2$. The former

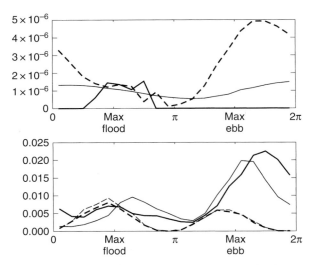

Fig. 4.9. Tidal cycles of mixing and dissipation rates. Top: depth-averaged mixing rates (kg m$^{-3}$ s$^{-1}$). Thin, diffusion $S_c = 0.1$; dashed, diffusion $S_c = 1$; thick, overturning. Bottom: dissipation rates (W m$^{-3}$). Full, internal shear, dashed, bed friction; thin $S_c = 1.0$, thick $S_c = 10$.

shows a pattern directly linked to tidal current speed. The vertical shear shows peak values on the decelerating phases of both ebb and flood flows but with markedly stronger values on the ebb. Interestingly, the peak values of dissipation, order of ($2 \times 10^{-2}$ W m$^{-3}$), are similar for both values of the $S_c = 10$ and 1. Rippeth *et al.* indicated comparable rates of dissipation up to $0.5 \times 10^{-2}$ W m$^{-3}$.

### *Tidal cycle and depth profiles*

Figure 4.10 (Prandle, 2004a) shows that tidal cycles of vertical differences for salinity for the Simulation 3 runs with $S_c = 10$ and 1.0. For $S_c = 1.0$, complete vertical mixing, due to convective overturning, is shown to start about 1.5 h after the start of the flood tide and to continue for about one-third of the cycle. Conversely, for $S_c = 10$, convective overturning only occurs at the end of the flood tide and is limited in extent to near-bed and near-surface.

### *4.4.5 Model applications for a range of estuarine conditions*

Following the above formulation and evaluation of the single-point model, a range of simulations were completed corresponding to

depths $D = 4, 5.7, 8, 11.3, 16, 22.6, 32, 45.3, 64$ m
density gradients $S_x = (\Delta \rho / \rho)/L_I$, $L_I = 12.5, 17.7, 25, 35.5, 50, 70.7, 100, 141.4, 200$ km
tidal velocity $U^* = 0.5$, bed friction $f = 0.0025$, Schmid number $S_c = 10$

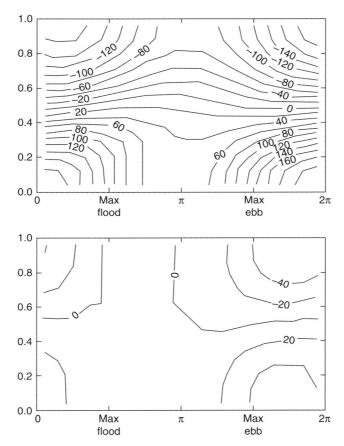

Fig. 4.10. Depth profiles of salinity differences over a tidal cycle. Top $S_c = 10$, bottom $S_c = 1.0$; contours 0.1‰.

$\Delta\rho = 0.027\rho$ is ocean salinity and $L_I$ the length of saline intrusion. Simulations were limited to conditions where the bed-to-surface salinity difference, $\delta s$, estimated from (4.40) was less than 10‰. Prandle (2004a) provides full details of these model simulations.

### Tidal velocities

Figure 3.3 indicates the form of the solution (3.16) for tidal current profiles. The difference in amplitude between surface and bed increases monotonically with larger values of the Strouhal number ($S_R = U^*P/D$), approaching an asymptote close to $S_R = 350$. Since Strouhal numbers in almost all meso- and macro-tidal estuaries will be well in excess of this value, the effects of tidal straining are likely to be broadly similar across a wide range of such estuaries. In simulations extending from $S_R = 200$ to 10 000, the present applications show a close fit between theory and model for both amplitude and phase of tidal currents. Moreover, these results for vertical profiles of tidal current amplitude and phase are largely insensitive to salinity gradients over the range considered. Hence, further consideration of tidal currents is omitted.

Souza and Simpson (1997) used H.F. Radar observations of surface tidal ellipses in a coastal (freshwater) plume region to indicate how a two-layer tidal response can develop. In the limit, the surface layer response is effectively frictionless, and the lower layer responds like a water column reduced in depth to that at their interface. While such results may be encountered in estuaries with weak tides, the associated dynamics involve convective and Coriolis terms omitted from (4.4).

### *Bed-to-surface residual current differences $\delta u$ and sea level gradient $\partial \varsigma / \partial X$*

Figure 4.11(a) (Prandle, 2004a) shows the values of $\delta u$, the bed-to-surface difference in tidally averaged velocities derived from (4.15) and Simulations 1 and 3 (using the model described in Section 4.4.3). The figure indicates values of $\delta u$ throughout the range of both intrusion lengths, $L_I$, and water depths, $D$, noted above. The results shown are for $U^* = 0.5\,\mathrm{m\,s^{-1}}, f = 0.0025$ and $S_c = 10$. The dots in Figs 4.11 and 4.12 represent values of $(D, L_I)$ from the eight observational data sets listed in Tables 4.2 and 4.3.

The use of axes with scales based on $\log_2(200/L_I)$ and $\log_2(D/4)$ results in contour distributions with slope $a = n/m$ for a parameter dependent on $L_I^m D^n$. Thus, for (4.15), where $\delta u \propto D^2/L_I\,fU^*$, the slope is $-2$ and the spacing is $\log_2 R^{1/n}$ on the $D$-axis and $\log_2 R^{1/m}$ on the $L_I$ axis, where $R$ is the ratio between contour values.

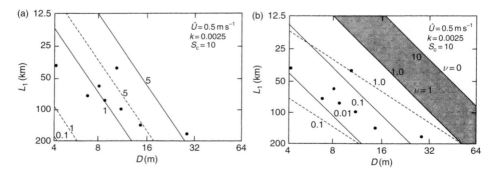

Fig. 4.11. Bed to surface changes as $f$ (depth, $D$, and saline intrusion length, $L_I$). (a) residual velocity (cm s$^{-1}$), (b) salinity (‰). Dashed lines, numerical model, full lines, theory (4.15) and (4.40). Dots indicate observed values (Prandle, 2004b). Shaded area in (b) indicates $0 < v < 1$, where $v$ (Hansen and Rattray, 1966), is the ratio of salt flux associated with vertical diffusion to that due to gravitational convection.

The values of $\delta u$ from Simulation 1 $(E=f\,U^{*}D$, no overturning) are in close agreement with the derived expression (4.15), as indicated previously. However, the introduction of overturning into Simulation 3 significantly enhances values of $\delta u$ relative to (4.15) while maintaining the theoretical dependence on $D^2/L_I$.

Associated results for the magnitude of the tidally averaged sea surface gradient $\partial\varsigma/\partial X$ from both Simulations 1 and 3 are in precise agreement with the derived values from (4.16). Over the ranges of $D$ and $L_I$ considered, $\partial\varsigma/\partial X$ is shown to increase up to $10 \times 10^{-6}$ which corresponds to a net increase in sea level of 1 cm km$^{-1}$ of intrusion length.

### *Bed-to-surface density differences, $\delta s$*

Figure 4.11(b) shows values of $\delta s$, the tidally averaged bed-to-surface difference in salinity. Again, results from Simulation 1 (without overturning) are in close agreement with the derived expression from (4.40), maintaining the dependency on $D^3/L_I^2$.

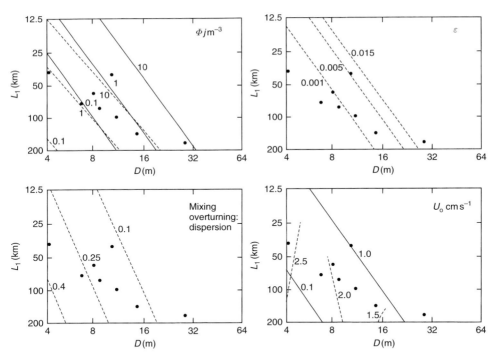

Fig. 4.12. Potential energy anomaly, $\varphi_M$; mixing efficiency, $\varepsilon$; overturning: diffusivity; river flow. Contours as for Fig. 4.11.
(Top left) Tidal mean potential energy anomaly $\varphi_M$, (4.64), in J m$^{-3}$.
(Top right) Efficiency of mixing, $\varepsilon$.
(Bottom left) Tidal mean ratio of mixing by convective overturning : vertical diffusivity.
(Bottom right) Balancing river flows, $U_0$ (cm s$^{-1}$).

However, as previously indicated in Fig. 4.8, results from Simulation 3 (with overturning) are radically different with enhanced values of $\delta s$. Moreover, these contours suggest a dependency on $D/L_I$. The tidal cycles shown in Fig. 4.10 indicate how overturning eliminates the balancing of ebb–flood tidal straining maintained in Simulation 1.

### 4.4.6 Mixing processes

The potential energy anomaly $\varphi_E$, is defined as

$$\varphi_E = \frac{1}{D} \int_0^D s' g (Z - D) \mathrm{d}Z, \tag{4.64}$$

where $s'$ is determined from the tidal current structure indicated in (4.40).

This represents the amount of energy required to mix the water column to a uniform density and hence, inversely, the effectiveness of vertical mixing.

Substituting (4.40) in (4.64), we obtain a time-averaged value of $\varphi_E$ associated with the tidally averaged density structure:

$$\phi_M = \frac{0.0007 \rho g^2 S_x^2 D^6}{E K_z}. \tag{4.65}$$

The tidally averaged results for $\varphi_M$ shown in Fig. 4.12 correspond to the density differences described in the previous section. For Simulation 1, without overturning, calculated values of $\varphi_M$ (not shown) are in close correspondence with (4.65) maintaining a dependency on $D^2/L_I$. Whereas, for Simulation 3 with overturning, radically enhanced values of $\varphi_M$ are obtained with a dependency closer to $D^{5/3}/L_I$. Typically, larger values of $\varphi_E$ are found over much of the tidal cycle, but these are reduced to zero by overturning over as much as one-third of the tidal cycle.

Details of the separate mixing processes within the simulations can be quantified. Figure 4.12 (Prandle, 2004a) shows, for Simulation 3, the ratio of (tidally averaged) mixing by convective overturning to that by vertical diffusivity. The maximum value of this ratio is 0.4 and occurs in shallow water for large values of $L_I$. This ratio decreases to 0.1 in deeper more stratified conditions.

The effectiveness of mixing, $\varepsilon$, is defined by Simpson and Bowers (1981) as the ratio of work done by mixing to that by tidal friction at the bed. Figure 4.12 shows, for Simulation 3, values of $\varepsilon$ ranging from less than 0.001 in well-mixed conditions up to 0.015 in more stratified conditions. Here, the total work done includes both that by bed friction ($\rho f U^3{}_{bed}$) and that by internal shear $\int \rho E_z (\mathrm{d}U/\mathrm{d}Z)^2 \, \mathrm{d}Z$. Interestingly, the Simpson and Bowers (1981) estimate of $\varepsilon \approx 0.004$ lies close to the centre of the calculated distributions. Hearn (1985) noted the ubiquity of

$\varepsilon \sim 0.003$ in shelf seas. He also shows values of $\varepsilon$ ranging from 0.003 to 0.016 based on observations at three sites (90–100 m deep) in the Celtic Sea.

Figure 4.12 also shows the value of $U_0$ required to balance the total rate of mixing. In Simulation 3, these values range from a maximum of $2.5 \, \text{cm s}^{-1}$ in shallow water to $1.5 \, \text{cm s}^{-1}$ in deeper water. A value of $3 \, \text{cm s}^{-1}$ is suggested from (4.58) due to tidally averaged tidal straining and between 0.1 and $1.0 \, \text{cm s}^{-1}$ from (4.59) due to vertical density differences. The latter values are closer to typical observed values shown in Tables 4.2 and 4.3 and suggest that some relaxation of constant $S_x$ may occur that reduces mixing associated with tidal straining.

## 4.5 Stratification

### 4.5.1 Flow ratio, $F_R$

It is generally accepted that stratification strengthens with increased values of depth $D$ and river flow $Q$ and with decreased values of breadth $B$ and tidal current amplitude $U^*$. The flow ratio, $F_R$, is defined as the net freshwater flow over a tidal cycle divided by the tidal prism, $T_P$, entering the estuary each flood tide, i.e. approximately the volume between high and low waters:

$$F_R = \frac{QP}{T_P} \sim \frac{Q\pi}{U^* BD} = \frac{U_0 \pi}{U^*}. \tag{4.66}$$

The above assumes that $T_P$ can be approximated by $U^* A \, P/\pi$, where $A = BD$ is the cross-sectional area. Schultz and Simmons (1957) suggested that estuaries are well-mixed for $F_R < 0.1$, i.e. $U_0 < 0.03 \, U^*$.

The parameter dependencies in (4.66) confirm the tendencies towards stratification noted above – with the exception of the influence of depth $D$. The following section explains this anomaly, illustrating how the kinetic relationship (4.66) needs to be extended to consider the balance between the rate of mixing (often concentrated near the surface in stratified flow) and the tidal dissipation (concentrated close to the bed). Chapter 3 shows that for $(\omega D/fU^*) > 0.25$, i.e. $D/U^* > 5$ s for a semi-diurnal tide, maximum velocities occur around mid-depth, with little current shear in the top half of the water. Subsequent sections confirm that, except for deeper micro-tidal estuaries where mixing scarcely extends to the surface, (4.66) is the primary indicator of stratification.

### 4.5.2 Energy and time requirements

The Simpson–Hunter (1974) criteria for stratified conditions is based on the ratio between the increase in potential energy due to mixing and the associated work done by tidal friction. The quantitative value indicative of mixing throughout the depth,

$$H/U^{*3} < 55 \text{ m}^2 \text{ s}^3, \tag{4.67}$$

is based on observational evidence of thermal stratification in shelf seas.

Prandle (1997) indicated that stratification levels can be calculated from the time, $D^2/K_z$, for complete vertical mixing by diffusion of a point source introduced at the surface or bed. Inserting (4.3) and specifying the time limit as the duration of the ebb or flood phases, stratification requires

$$\frac{D^2}{K_z} = \frac{D}{S_c f U^*} > 6 \text{ h} \quad \text{or} \quad \frac{D}{U^*} > 56 \ S_c \text{ s}. \tag{4.68}$$

Since $U^*$ is usually in the range $0.5$–$1.0 \text{ m s}^{-1}$, and $S_c$ between $0.1$ and $1$, we see the correspondence with the Simpson–Hunter criterion. Most estuaries have values of $D^2/K_z > 1$ h, indicating that some degree of intra-tidal stratification will occur. In Chapter 6, it is shown how, for a synchronous estuary, both criteria correspond closely to stratification occurring for tidal elevation amplitudes $\varsigma^* \leq 1$ m.

### 4.5.3  Richardson number

Time- and depth-averaged values for $R_i$ (4.2) can be calculated using (4.40) for $\partial \rho / \partial Z$ and (4.55) for $\partial U / \partial Z$, giving

$$R_i = \frac{\dfrac{68 g S_x^2 D^4}{10^4 E K_z}}{\left(0.45 \dfrac{U^*}{D}\right)^2} = 100 \ S_c \left(\frac{U_0}{U^*}\right)^2. \tag{4.69}$$

Thus, with $E$ from (4.3), $S_x = 0.027/L_I$, where $L_I$ is the saline intrusion length from (4.44). The condition, $R_i < 0.25$, for mixing to occur, corresponds to $U_0 < 0.05 \ U^*$ for $S_c = 1$ and $U_0 < 0.016 \ U^*$ for $S_c = 10$.

### 4.5.4  Stratification number

Ippen and Harleman (1961) demonstrated that vertical mixing could be related to the balance between turbulence associated with the rate of dissipation of tidal energy by bed stress, $G = \varepsilon (4/3\pi) f \rho U^{*3} L_I$, and the energy required to increase the potential energy level by vertical mixing, $J = \frac{1}{2} \Delta \rho \ g \ H^2 U_0$. The parameter $\varepsilon$ representing the efficiency of mixing, typically $\sim 0.004$ (Section 4.4.6), is introduced here to give a modified stratification number, defined as

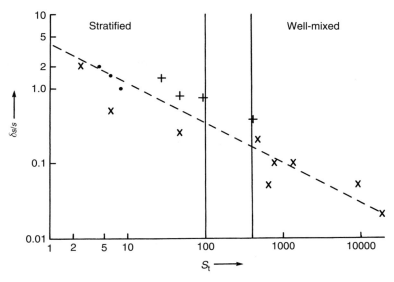

Fig. 4.13. Level of stratification $\delta s/s$ versus Stratification number $S_T$. $S_T = 0.017$ $\varepsilon (U^*/U_0)^2$, (4.70). • Rigter's (1973) flume tests, + WES flume tests, × observations.

$$S_T' = S_T \varepsilon = \frac{\varepsilon G}{J} = 0.85 \frac{\varepsilon f\, U^* L_I}{\frac{\Delta\rho}{\rho} g H^2 U_0} = 0.017\, \varepsilon \left(\frac{U^*}{U_0}\right)^2. \qquad (4.70)$$

Thus, the stratification number extends the concept of the Simpson–Hunter criterion described in Section 4.5.2 to balance net mixing of fresh water input over the saline intrusion length. Prandle (1985) showed that estuaries change from mixed to stratified as the stratification number decreases from above 400 to below 100 (Fig. 4.13; Prandle, 1985). Adopting the limit $S_T = 250$, then for $\varepsilon = 0.004$, the boundary corresponds to $S_T' = 1$ and $U_0 < 0.01\, U^*$.

### 4.5.5 Vertical salinity difference

Hansen and Rattray (1966) indicated that a normalised surface to bed salinity differences of $\delta s/s < 0.1$ could be adopted as the boundary for mixed and stratified estuaries. From Fig. 4.13 and (4.70),

$$\frac{\delta s}{s} = 4\, S_T^{-0.5} = 31 \left(\frac{U_0}{U^*}\right). \qquad (4.71)$$

Thus, $\delta s/s < 0.1$ implies that estuaries will be mixed for $U_0 < 0.003\, U^*$. Using the suggested revised demarcation, based on Fig. 4.13, of $S_T = 250$, stratification is defined as $\delta s/s > 0.25$, $U_0 > 0.01\, U^*$.

## 4.6  Summary of results and guidelines for application

Within estuaries, saline intrusion can impact significantly on flora, fauna, biology, chemistry, sediments, morphology, flushing time, contaminant pathways, etc. Here we examine the dynamics of mixing, indicating how bathymetry and both marine and fluvial conditions determine the degree of stratification and the extent of axial penetration. Appendix 4A outlines the characteristics of the related dependency of water density on temperature, illustrating how the seasonal cycle varies according to the level of stratification and latitude.

The leading question is:

***How does salt water intrude and mix and how does this change over cycles of spring–neap tides and flood-to-drought river flows?***

Apart from an occasional surface scum line, the ebb and flood of saline intrusion in estuaries passes largely unnoticed. Yet, the extent of saline intrusion was often the determining factor in the siting of towns or industries reliant on fresh river water. Moreover, the extent and nature of intrusions determines net estuarine concentrations of both dissolved marine tracers and fluvial contaminants. At the interface with river water, salt water produces electrolytic attraction between fine suspended sediments resulting in rapidly settling 'flocs' which accumulate at the seaward (ebb) and landward (flood) limits of intrusion.

In strongly tidal estuaries, saline intrusion has little impact on tidal propagation (Prandle, 2004a). Conversely, the nature of saline intrusion is overwhelmingly determined by the combination of tidal motions alongside the flow of river water. Pritchard (1955) introduced a generalised classification of estuaries according to their salinity intrusion, varying from fully mixed (vertically) in strongly tidal, shallow estuaries with small river flows through to 'arrested saline wedge' in deeper, micro-tidal estuaries with large river flows. Hansen and Rattray (1966) produced a generalised stratification diagram (Fig. 4.3), converted by Prandle (1985) in terms of more readily available parameters. However, this diagram does not directly answer the above question concerning the range of saline intrusion in estuaries.

Stratification is complicated by continuous interactions between the dynamics promoting mixing and associated adjustments to the background salinity gradients. The Richardson number (4.2), quantifies the ratio of 'stabilising' buoyancy inflow to the rate of turbulent mixing. It has proven a robust indicator of stratification across a wide range of fluid dynamics. However, this number can vary appreciably over the tidal and fluvial cycles noted above and from mouth to head and laterally within an estuary.

Section 4.3 describes how various approaches for estimating the length of saline intrusion, $L_I$, all suggest a dependence on $D^2/f\ U^*U_0$ ($D$ depth $f$ bed friction coefficient, $U^*$ tidal current amplitude and $U_0$ residual current associated with river flow). Assessments against a range of flume tests indicate the validity and robustness of this formulation. However, this formula fails to account for observed variations in intrusion lengths (Uncles *et al.*, 2002). The additional factor, which must be incorporated in funnel-shaped estuaries, is axial migration of the intrusion. This introduces a complex inter-dependency between the length and the location of the intrusion. Analysis of observations suggests that this axial migration adjusts to enable mixing to occur as far seawards as is consistent with containing the mixing within the estuary. The time delay for such axial migration of the intrusion is related to the estuarine flushing time $T_F$. It is shown in Chapter 6 and from observations that $T_F$ is in the range of days, and hence changing tidal and river flow conditions over such periods further complicates the calculation of $L_I$. Clearly, model simulations need to extend adequately in time to accommodate such migrations and incorporate appropriate 'relaxation' of both seaward and landward boundary conditions.

Abraham (1988) noted the significance of 'tidal straining', whereby on the flood tide, larger near-surface tidal velocities advect denser more saline water over fresher lower layers leading to mixing by convective overturning. Simpson *et al.* (1990) provided both theoretical and observational quantification of this phenomenon. However, despite the neglect of this latter effect, 'tidally averaged' analytical solutions for salinity-induced vertical profiles of residual velocities and salinities are found to be widely applicable. The summary of these analytical solutions in Section 4.2.6 and Table 4.1 emphasises how, for steady-state equilibria, the dynamical adjustments to saline intrusion involve small (relative to tidal components), residual currents and surface gradients.

These solutions provide estimates for residual flow components associated with (i) river flow $Q$ (4.12), (ii) wind stress $\tau_w$ (4.37), (iii) a well-mixed longitudinal density gradient (4.15), and (iv) a fully stratified saline wedge (4.32) and (4.33). The relative magnitudes of each of these components are defined in terms of dimensionless parameters (4.42). The residual current profiles (i), (ii) and (iii) are shown in Fig. 4.4.

From Table 4.1, the density forcing term $0.5S$ is balanced by a surface gradient $-0.46S$ with the remaining component driving a residual circulation of $0.036S$ $U_0/F$ seawards at the surface and $-0.029S$ $U_0/F$ (landwards) at the bed. Similarly, the wind stress term $-W$ is counteracted by a surface gradient $1.15W$ with the 'excess' balance driving a circulation of $0.31W$ $U_0/F$ at the surface and $-0.12W$ $U_0/F$ at the bed ($W$, $S$ and $F$ are defined in (4.42)). Thus, for steady-state conditions, both wind and density forcing are mainly balanced by surface gradients with only a small fraction of the forcing effective in maintaining a vertical circulation.

Section 4.4 uses these analytically derived current and salinity profiles to estimate the (localised) rates of vertical mixing and compares these alongside the values associated with tidal straining to the rate of supply of river water. These analytical solutions are further compared with results from a 'single-point' numerical model which incorporates tidal straining and convective overturning whenever density decreases with depth. The model also includes the Munk and Anderson (1948) modifications of the eddy diffusivity and eddy viscosity coefficients, $E$ and $K_z$, based on Richardson number representations of buoyancy effects.

The application of this numerical model extends to a range of values of both intrusion lengths, $L_I$, and water depths, $D$. Sensitivities to the Schmidt number, expressing the ratio of $E:K_z$ are also explored. Results are shown for surface-to-bed differences in residual velocity, $\delta u$, and salinity, $\delta s$; potential energy anomaly, $\varphi_E$; the ratio of mixing by diffusivity: overturning; the efficiency of mixing $\varepsilon$ and residual river flow $U_0$ (required to balance mixing rates).

Qualitatively, both numerical model and analytical solutions for the residual currents and salinity profiles are in close agreement. However, the model suggests that tidal straining can increase the magnitudes of these currents and salinity structure several fold. Averaged over a tidal cycle, the component of mixing associated with overturning is shown to be generally significantly less than mixing by diffusion, typically $0.1-0.4$ of the latter. However, the pronounced impact of overturning in removing stratification on the flood tide is evident. The model also quantifies mixing rates related to bed friction and internal shear. The results confirm Simpson and Bower's (1981) observational findings that less than 1% of the energy involved in tidal dissipation is effective in promoting vertical mixing.

Section 4.5 revisits the issue of indicators for estuarine stratification. New theories on stratification are reconciled with historical indicators, emphasising that $U_0/U^* > 0.01$ is the common key indicator of stratification. Calculations are made of likely values of $U_0$ in the saline intrusion zone based on (i) most seaward location of mixing (4.52); (ii) a flow ratio, $F_R = U_0 \pi/U^* = 0.1$ (4.66); (iii) a Richardson number $R_I = 0.25$ (4.69); (iv) a balance between gain in potential energy and tidal dissipation (4.70); and (v) observations of stratification (Fig. 4.13). All five approaches indicate values of $U_0$ close to 1 cm s$^{-1}$ in the intrusion zone of 'mixed' estuaries – a result subsequently further confirmed in Chapter 6. This indicator confirms that stratification generally increases with larger river flow, narrower breadths and weaker tidal currents. However, the indication that stratification increases with shallower water is counteracted by reduced tidal current shear nearer the surface in deep water and the related reduced role of tidal straining.

Appendix 4A provides expressions for the mean and seasonal variations of both the water surface and the ambient air as functions of depth, latitude and degree of stratification.

## Appendix 4A

### *4A.1 The annual temperature cycle*

A generalised theory is developed to describe the annual temperature cycle in estuaries and adjacent seas. A sinusoidal approximation to the annual solar heating component, $S$, is assumed and the surface loss term is expressed as a constant, $k$, times the air-sea temperature difference $(T_a - T_s)$. For vertically mixed conditions, analytical solutions show that in shallow depths, the water temperature follows closely that of the ambient air temperature with limited separate effect of solar heating. Conversely in greater depths, the water surface temperature variations will be reduced relative to those of the ambient air. Providing such deep water remains mixed vertically, the seasonal variation will be inversely proportional to depth and maximum temperatures will occur up to 3 months after the maximum of solar heating. The annual mean water temperature will exceed the annual mean air temperature by the annual mean of $S$ divided by $k$. For wider applications, a numerical model is used to derive generalised expressions for the mean and amplitude of both air and water surface temperatures as functions of latitude, depth and tidal current speed.

The description of temperature in marine ecological models is important both directly for its influence on specific parameters and indirectly via its contribution to vertical density variations.

Water density can be approximated by $\rho = 1000 + 0.7S - 0.2\,T\,(\mathrm{kg\,m^{-3}})$, where $S$ is salinity in ‰ and $T$ is temperature in °C. Thus, stratification at the demarcation limit, $\delta S \sim 0.25‰$, is equivalent to a surface-to-bed temperature difference of $0.875\,°\mathrm{C}$ .

Simplified and generalised approaches are used to examine how the seasonal temperatures vary with (i) the level of solar heating (i.e. cloud cover and latitude), (ii) ambient air temperature, (iii) wind speed (strongly controlling the rate of air-sea heat exchange), (iv) water depth and (v) the degree of vertical stirring. Results from theses studies (Prandle and Lane, 1995; Prandle, 1998) can be used to examine the impact of changes in any of these parameters that might arise from climate trends, etc.

The fundamental simplifications introduced are as follows:

(1) approximation of the annual solar heat input by a mean value plus a sinusoidal term: $S_0 - S^* \cos \omega t$;
(2) representation of the heat losses by a term $k(T_s - T_a)$, where $T_s$ is the water surface temperature, $T_a$ the air temperature and $k$ a constant coefficient;
(3) assumption of a localised equilibrium, i.e. neglect of horizontal components of advection and dispersion;
(4) representation of vertical mixing processes by an eddy dispersion coefficient encompassing the effects of vertical mixing processes by both vertical advection and dispersion;
(5) neglect of both the (solar) diurnal and the neap–spring tidal cycles of variability.

The adoption of a sinusoidal approximation to the annual temperature cycle involves phase values with 360° corresponding to 1 year. For convenience, the 'year' is here shortened to 360 days so that 1° corresponds to 1 day. Where 'day numbers' are cited, strictly, these start at the winter solstice.

### 4A.2  Components of heat exchange at the sea surface

Goldsmith and Bunker (1979) describe four components involved in the heat exchange between the atmosphere and the sea, namely

(1) *SR* – solar radiation, i.e. the primary energy source;
(2) *LH* – latent heat flux, heat lost from the sea by evaporation (occasionally gained by condensation);
(3) *IR* – infrared back radiation, the net effective 'reflection' from the sea;
(4) *LS* – sensible heat flux, heat exchange between air and sea by 'conduction'.

Figure 4A.1 shows the annual variation of *SR*, *LH*, *IR* and *LS* at 55°N, 3°E (northern North Sea) based on data for 1989 supplied by the UK. Meteorological Office, Cave (1990) and heat exchange terms calculated from Goldsmith and Bunker (1979). The effective value of *SR* at the sea surface is reduced by cloud cover, this reduction is about 1/8 for 25% cover, 1/3 for 50% cover and 4/5 for 100% cover. However, the annual cycle (frequency $\omega$) at almost all latitudes can be approximated by

$$SR = S(1 - \beta \cos \omega t). \tag{4A.1}$$

The latent heat flux *LH* is a function of wind speed *W*, the air–sea temperature difference $(T_a - T_s)$ and relative humidity.

The effective back radiation *IR* depends on the absorptive properties of the atmosphere and both $T_a$ and $T_s$; however, it is most sensitive to $T_a$.

The sensible heat flux *LS* is proportional to both wind speed *W* and $(T_s - T_a)$. Air gains heat quicker than water as a consequence of their relative thermal conductivities, i.e. 0.0226 and 0.60 W m$^{-1}$ °C$^{-1}$, respectively, and hence the assumption of a constant value for *k* may be invalid in seas where $T_a > T_s$ for extended periods.

For simplicity, we lump together solar radiation *SR* and infrared back radiation *IR* into a single heat gain term *S* approximated by

$$S(t) = S_0 - S^* \cos \omega t. \tag{4A.2}$$

Likewise, we lump together the *LH* and *LS* into the single heat loss term *L* where

$$L(t) = -k(T_s - T_a). \tag{4A.3}$$

Annual mean solar radiation, $S_0$ in (4A.2) in W m$^{-2}$, in the absence of cloud cover, ranges from 317 at the equator to 234 at 45° latitude and 133 at the poles. In

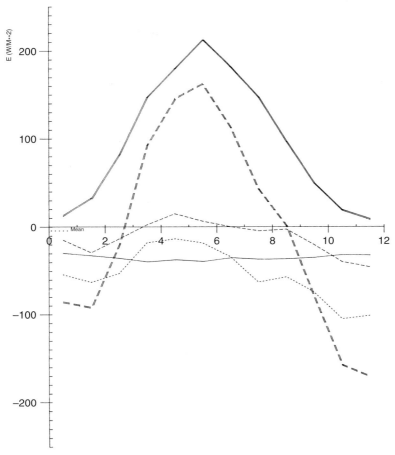

Fig. 4A.1. Annual cycle of *SR*, == solar radiation; – – – – *LH*, latent heat flux; ——
*IR*, infrared back radiation and *LS*, ---- sensible heat flux, === net. Values for 1989
at positions (55° N, 3° E) in the North Sea (Units $W\,m^{-2}$).

Corresponding values for $S^*$ range from 0 at the equator, to 147 at 45° and 209 at the
poles. A second harmonic, with frequency $2\omega$, amounts to 16 W m$^{-2}$ at 60°N, rising
to 41 at 70° and 87 at 90°. Thus, the present theoretical approach which assumes just
the first harmonic in (4A.1) is reasonable in latitudes up to 70°. Fig. 4A.1 (Prandle
and Lane, 1995) shows a typical annual cycle in the North Sea.

### 4A.3  *Analytical solutions in vertically mixed water*

From the assumptions (4A.2) and (4A.3), the rate of change of the sea temperature
$T_s$ is

$$\frac{\partial Ts}{\partial t} = \frac{S(t) + L(T)}{\alpha D},\tag{4A.4}$$

where $\alpha$ is the thermal capacity of water (= $3.9 \times 10^6 \, \mathrm{J \, m^{-3} \, {}^\circ C^{-1}}$).
  Assuming

$$T_a = \bar{T}_a - \hat{T}_a \cos(\omega t - g_a),\tag{4A.5}$$

since the equation is linear, $T_S$ must take the form

$$T_s = \bar{T}_s - \hat{T}_s \cos(\omega t - g_s),\tag{4A.6}$$

where $g_a$ and $g_s$ are the phase lags of air and sea surface temperature relative to the solar radiation cycle (4A.1).
  Substituting (4A.1), (4A.3), (4A.5) and (4A.6) into (4A.4) gives

$$\begin{aligned}\alpha D\omega \, \hat{T}_s \sin(\omega t - g_s) = \; & S_0 - S^* \cos \omega t \\ & + k(\bar{T}_a - \hat{T}_a \cos(\omega t - g_a) - \bar{T}_s + \hat{T}_s \cos(\omega t - g_a)).\end{aligned}\tag{4A.7}$$

Thus, for the time invariant terms,

$$\bar{T}_s = \bar{T}_a + \frac{S_0}{k}\tag{4A.8}$$

and for the seasonal cycle,

$$\hat{T}_s \cos(\omega t - g_s - B) = \frac{S^* \cos \omega t + k\hat{T}_a \cos(\omega t - g_a)}{(k^2 + \alpha^2 D^2 \omega^2)^{1/2}},\tag{4A.9}$$

where $B = \arctan(-\alpha D\omega, k)$.
  Figure A4.2 (Prandle and Lane, 1995) shows the values of $\hat{T}_s / \hat{T}_a$ for a range of values of $\hat{T}_a$ and $D$. These results correspond to $S_0 = S^* = 100 \, \mathrm{W \, m^{-2}}$ and $k = 50 \, \mathrm{W \, m^{-2} \, {}^\circ C^{-1}}$ with $g_a = 30°$. These results are essentially similar for $0° < g_a < 90°$ and for either $S^*$ or $k$ changed by a factor of 2.
  Referring to (4A.9), four quadrants can be distinguished in Fig. 4A.2 differentiating shallow and deep water and small and large $\hat{T}_a$.

In quadrant **Q1**, $D \ll k/\alpha\omega$ and $\hat{T}_a \gg S^*/k$, thus $\hat{T}_s \to \hat{T}_a$ and $g_s \to g_a$.
In quadrant **Q2**, $D \gg k/\alpha\omega$ and $\hat{T}_a \gg S^*/k$, thus $\hat{T}_s \to \hat{T}_a k/D\alpha\omega$ and $g_s \to g_a + 90°$.
In quadrant **Q3**, $D \ll k/\alpha\omega$ and $\hat{T}_a \ll S^*/k$, thus $\hat{T}_s \to S^*/k$ and $g_s \to 0°$.
In quadrant **Q4**, $D \gg k/\alpha\omega$ and $\hat{T}_a \ll S^*/k$, thus $\hat{T}_s \to S^*/D\alpha\omega$ and $g_s \to 90°$.

### 4A.4  Coupled atmosphere-marine model

A major limitation in using the above analytical results for sensitivity tests arises from the need to prescribe a (fixed) air temperature, thus precluding the feedback

(a)

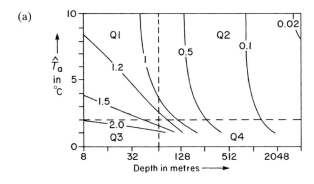

(b)

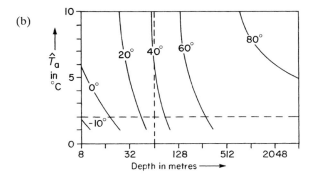

Fig. 4A.2. The annual temperature cycle in well-mixed waters. (a) Top $\hat{T}_\mathrm{s}/\hat{T}_\mathrm{a}$ (b) bottom $g_\mathrm{s}$–$g_\mathrm{a}$ as functions of $\hat{T}_\mathrm{a}$ and depth results correspond to $S_0=S^*=$ 100 W m$^{-2}$, $k=50$ W m$^{-2}$ °C$^{-1}$, $g_\mathrm{a}=30$ °C.

impact of changes in sea temperature on ambient air temperature. Clearly, there is a close thermal coupling between the atmosphere and the sea with characteristic cycles at daily, monthly and semi-monthly (spring–neap tides) and annual frequencies. Prandle (1998) formulated a 'single point' coupled air–sea thermal exchange model concentrating on reproduction of the annual temperature cycle of the sea surface and ambient air. The following is a brief description of this model.

A marine model is linked to an atmospheric model with air–sea exchange rates from Gill (1982), comprising long wave radiation, evaporation and convection (or sensible heat flux). Vertical exchange within the water column is governed by both tidal- and wind-forced turbulent intensity levels, modulated by vertical density gradients. Incident solar radiation is modulated by a reflection coefficient at the outer edge of the atmosphere and thence by internal absorption. The heat lost by the sea surface is subsequently absorbed by the atmosphere, except for 0.3 $Q_\mathrm{L}$, which is assumed to radiate directly to space. The atmospheric model effectively forms a 'surface layer' to the marine model with external boundary conditions (incident

Table 4A.1 *Model parameters*

| | |
|---|---|
| Atmospheric reflection | $r = -0.47 + 0.86 \cos \lambda^{1/2}$ ($\lambda$ latitude) |
| Atmospheric absorption coefficient | $A = 0.11$ |
| Atmospheric temperature gradient | $\Delta T = 42.5 \, °C$ |
| Cloud cover | $C = 0.5$ |
| Minimum eddy diffusivity | $K_z = 10^{-5} \, (\text{m}^2 \, \text{s}^{-1})$ |
| Atmospheric height (water equivalent) | $d = 2.5 \, (\text{m})$ |
| Relative humidity | $R = 0.8$ |
| Wind speed (m s$^{-1}$) | $W = 6(1 + (\lambda/65)^2)(1 + 0.5 \cos\omega_a t)$ |
| Number vertical grid | $n = 10$ to $100$ |
| Time step | $\Delta t = 900$s |

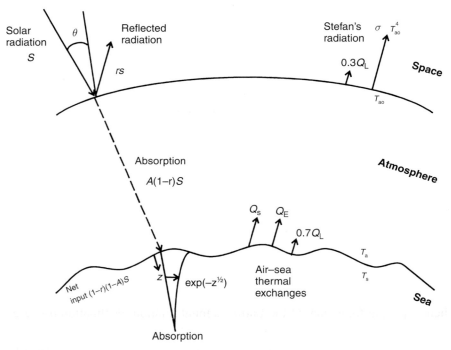

Fig. 4A.3. Schematic representation of atmosphere–marine thermal exchange model.

solar energy, radiated and reflected thermal energy) specified at the outer edge of the atmosphere. The external heat loss rate from the atmosphere follows Stefan's law. The salient features are indicated in Fig. 4A.3 and Table 4A.1 (Prandle, 1998).

The model parameters, listed in Table 4A.1, were adjusted to reproduce seasonal temperature cycles of the sea surface and ambient air observed in the North Atlantic, extracted by Isemer and Hasse (1983) over the latitude range 0–65°N. In shallow water (<200 m), the amplitude of this seasonal cycle is modulated by both the water depth and the tidal current amplitude with large tidal currents decreasing seasonal amplitudes.

*The marine model*

Temperatures within the water column are calculated from the vertical dispersion equation:

$$\frac{\partial}{\partial t} T = \frac{\partial}{\partial Z} K_Z \frac{\partial T}{\partial Z} + \frac{Q(Z)}{\alpha}, \tag{4A.10}$$

where $T$ is temperature, $K_Z$ vertical eddy diffusivity, $Q(Z)$ the atmospheric thermal input per unit depth at level $Z$, $\alpha$ the thermal capacity of water, $t$ time and $Z$ the vertical axis. The determination of $K_Z$ is via the Mellor–Yamada (1974) level 2.5 turbulence closure model (Appendix 3B). The tidal- and wind-driven currents are calculated from a solution of the momentum equation (Chapter 3).

### 4A.5  The atmospheric model – solar energy input, reflection, absorption and radiation

Using the notation from Fig. 4A.3, the net solar radiation into the sea is

$$SR = S(\cos\theta\,(1-r) - A), \tag{4A.11}$$

where $S = 1353\ \mathrm{W\,m^2}$ and the inclination of the sun relative to the vertical is given by

$$\cos\theta = \sin\beta \cdot \sin\lambda - \cos\beta \cdot \cos\lambda \cdot \cos(\chi + \omega_d t), \tag{4A.12}$$

where $\omega_d$ is the frequency of the Earth's daily rotation, $\lambda$ latitude and $\chi$ longitude:

$$\sin\beta = \sin\delta \cdot \sin(\omega_a t - \chi_o), \tag{4A.13}$$

where $\omega_a$ is the frequency of the Earth's annual rotation, declination $\delta = 23.5°$ and the 'reference' longitude $\chi_0 \sim 80°$ for $t$ measured relative to January 1.

The following expression was found necessary to reproduce observed (North Atlantic) air and sea surface temperatures:

$$r = -0.47 + 0.86(\cos\lambda)^{1/2}. \tag{4A.14}$$

The global average of solar energy per unit surface area is equal to 0.25 $S$ as computed directly from the ratio of aspect $\pi R^2$ to surface area $4\pi R^2$. The global mean value for the present formulation for the atmospheric reflection factor, $r$, is 0.30, in precise agreement with Gill (1982). The formulation for the atmospheric absorption coefficient, $A = 0.11$, corresponds to a global mean absorption factor of 0.15 while Gill estimated 0.19. Some flexibility exists between the respective formulations of $A$ and $r$. A latitudinal variability in $A$ could be introduced in the same manner as specified for $r$. A fixed temperature difference, $\Delta T$, of $42.5\,°C$ is assumed between the sea surface and the outer edge of the atmosphere. The computed mean sea surface temperature depends primarily on the prescriptions of $A$, $r$ and $\Delta T$.

### 4A.6 Global expressions

The following generalised expressions were derived from least squares fitting to model results:

$$\bar{T}_s = 40\cos\lambda - 12.5$$

$$\bar{T}_a = 35\cos\lambda - 10.0$$

$$\hat{T}_s = \frac{0.080\,\lambda}{1 - \exp(-D/50)} \quad \text{for } U^* < 0.2\,\mathrm{m\,s^{-1}}$$

$$\hat{T}_s = \frac{0.064\,\lambda}{1 - \exp(-D/50)} \quad \text{for } U^* > 0.2\,\mathrm{m\,s^{-1}}$$

$$\hat{T}_a = \frac{0.086\,\lambda}{1 - \exp(-D/50)} \quad \text{for } U^* < 0.2\,\mathrm{m\,s^{-1}}$$

$$\hat{T}_a = \frac{0.067\,\lambda}{1 - \exp(-D/50)} \quad \text{for } U^* > 0.2\ \mathrm{m\ s^{-1}}. \qquad (4A.15)$$

Temperatures in degrees centigrade, depths $D$ in metres and $U^*$ is the tidal current amplitude.

Figure 4A.4 (Prandle, 1998) shows salient results for the annual mean temperatures, $\bar{T}_s$ and $\bar{T}_a$. Mean temperatures are overwhelmingly determined by latitude, with little influence of water depth or tidal current amplitude.

Figure 4A.5 (Prandle, 1998) shows the values for seasonal amplitudes, $\hat{T}_s$ and $\hat{T}_a$, illustrating the dependence on latitude in combination with an exponential function of depth. The latter dependency increases $\hat{T}_s$ and $\hat{T}_a$ dramatically in shallow water. This dependency on water depth reaches an asymptotic limit for $D \gg 100$ m. In water depths $D > 40$ m, the values of both $\hat{T}_s$ and $\hat{T}_a$ increase significantly for the smallest tidal current amplitude $U^* = 0.1\,\mathrm{m\,s^{-1}}$. However, for $U^* > 0.2\,\mathrm{m\,s^{-1}}$, the values of both $\hat{T}_s$ and $\hat{T}_a$ converge asymptotically towards the reduced values shown in Fig. 4A.5.

The phases for the seasonal cycles indicate that maximum northern hemisphere sea surface temperatures almost always occur between Julian days

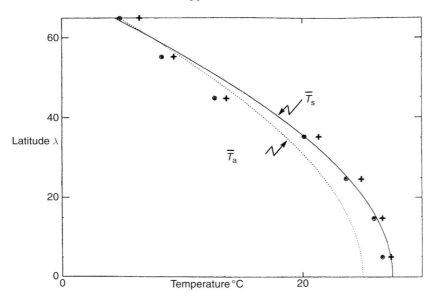

Fig. 4A.4. Computed annual mean temperatures: $\bar{T}_\mathrm{s}$ sea surface, $\bar{T}_\mathrm{a}$ ambient air (4A.15) Bunker Atlas Observations: $\otimes$ air, + sea surface.

230 and 255 – generally earlier in deeper water. The phases for air temperatures generally follow those of the sea surface with maximum temperatures typically 5–10 days later.

The coupled model is applicable for latitudes between 0°and 65°N; areas further polewards introduce particular difficulties with surface icing and with anomalous temperature–density functions. The model can be used for small-amplitude sensitivity analyses to examine, e.g. the impact of changes in cloud cover, wind speed, or relative humidity or the magnitude and persistence of the impact of a single major storm. The generality of the derived expressions for $T_\mathrm{s}$ and $T_\mathrm{a}$ provides a simple understanding of likely conditions over a range of varying latitudes and with varying water depths and tidal current amplitudes. Likewise, the model may be usefully adopted for interdisciplinary studies including feedback mechanisms whereby biological and chemical parameters impact on absorption and reflection coefficients.

In summary, the coupled ocean–atmosphere numerical simulations show that the mean values of both air and water temperatures are overwhelmingly determined by (the cosine of) latitude, with little influence of water depth or tidal current amplitude. By contrast, corresponding seasonal amplitudes vary directly with latitude alongside an exponential function of depth with much larger values in shallow weakly mixed waters. Increased stratification insulates the sea, especially at greater depths, from both solar heating and surface heat exchanges. The effect is to lower

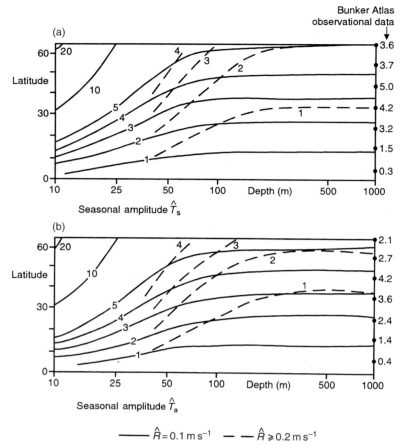

Fig. 4A.5. Seasonal amplitudes for (a) sea-suface temperature $\hat{T}_s$ (b) ambient air $\hat{T}_a$ (4A.15). Dashed contours are for tidal current amplitudes $U^* \geq 0.2\,\mathrm{m\,s^{-1}}$. Continuous contours are for $U^* < 0.2\,\mathrm{m\,s^{-1}}$. Digital values on the right-hand axis indicate observed values from the Bunker Atlas.

both the mean and the variability of deeper water temperatures, especially when autumnal overturning occurs.

# References

Abraham, C., 1988. Turbulence and mixing in stratified tidal flows. In: Dronkers, J. and vanLeussen, W. (eds), *Physical Processes in Estuaries*. Springer Verlag, Berlin, pp. 149–180.

Bowden, K.F., 1953. Note on wind drift in a channel in the presence of tidal currents. *Proceedings of the Royal Society of London*, A, **219**, 426–446.

Bowden, K.F., 1981. Turbulent mixing in estuaries. *Ocean Management*, **6** (2–3), 117–135.

Bowden, K.F. and Fairbairn, L.A., 1952. A determination of the frictional forces in a tidal current. *Proceedings of the Royal Society of London*, Series A, **214**, 371–392.

Cave W.R., 1990. *Re format Procedures and Software for Meteorological Office Data*. Unpublished Report. British Oceanographic Data Centre, Bidston Observatory, Birkenhead, UK.

Dyer, K.R. and New, A.L., 1986. Intermittency in estuarine mixing. In: Wolfe, D.A. (ed.), *Estuarine Variability. Proceedings of the Eighth Biennial International Estuarine Research Conference*, University of New Hampshire, Durham, 28 July–2 August, 1985. Academic Press, Orlando, pp. 321–339.

Geyer, W.R. and Smith, J.D., 1987. Shear instability in a highly stratified estuary. *Journal of Physical Oceanography*, **17** (10), 1668–1679.

Gill A.E., 1982. *Atmosphere–Ocean Dynamics*. Academic Press, Oxford.

Godin, G., 1985. Modification of river tides by the discharge. *Journal of Waterway Port Coastal and Ocean Engineering*, **111** (2), 257–274.

Goldsmith R.A. and Bunker, A.F., 1979. *WHOI Collection of Climatology and Air–Sea Interaction (CASI) Data*, Technical Report.

Hansen, D.V. and Rattray, M.J., 1966. New dimensions in estuary classification. *Limonology and Oceanography*, **11**, 319–326.

Hearn, C.J., 1985. On the value of the mixing efficiency in the Simpson–Hunter H/U$^3$ criterion. *Deutsche Hydrographische Zeitschrift*, **38**, 133–145.

Ippen, A.T. (ed.), 1966. *Estuary and Coastline Hydrodynamics*. McGraw-Hill, New York.

Ippen, A.T. and Harleman, D.R. F., 1961. One-dimensional analysis of salinity intrusion in estuaries. *Technical Bulletin* No. 5, Committee on Tidal Hydraulics Waterways Experiment Station, Vicksburg, MS.

Isemer H.J. and Hasse, L., 1983. *The Bunker Climate Atlas of the North Atlantic Ocean*, Vol. 1, Observations. Springer-Verlag, p. 218.

Lewis, R., 1997. *Dispersion in Estuaries and Coastal Waters*. John Wiley and Sons, Chichester.

Linden, P.P. and Simpson, J.E., 1988. Modulated mixing and frontogenesis in shallow seas and estuaries. *Continental Shelf Research*, **8** (10), 1107–1127.

Liu, W.C., Chen, W.B., Kuo, J-T. and Wu, C., 2008. Numerical determination of residence time and age in a partially mixed estuary using a three-dimensional hydrodynamic model. *Continental Shelf Research*, **28** (8), 1068–1088.

Mellor O.L. and Yamada, T., 1974. A hierarchy of turbulence closure models for planetary boundary layers. *Journal of the Atmospheric Science*, **31** (7) 1791–1806.

Munk, W.H. and Anderson, E.R., 1948. Notes on a theory of the thermocline. *Journal of Marine Research*, **7**, 276–295.

New, A.L., Dyer, K.R. and Lewis, R.E., 1987. Internal waves and intense mixing periods in a partially stratified estuary. *Estuarine, Coastal and Shelf Science*, **24** (1), 15–34.

Nunes, R.A. and Lennon, G.W., 1986. Physical property distributions and seasonal trends in Spencer Gulf, South Australia: an inverse estuary. *Australian Journal of Marine and Freshwater Research*, **37** (1), 39–53.

Nunes Vaz, R.A. and Simpson, J.H., 1994. Turbulence closure modelling of estuarine stratification. *Journal of Geophysical Research*, **95** (C8), 16143–16160.

Oey, L.Y., 1984. On steady salinity distribution and circulation in partially mixed and well mixed estuaries. *Journal of Physical Oceanography*, **14** (3), 629–645.

Officer, C.B., 1976. *Physical Oceanography of Estuaries*. John Wiley and Sons, New York.

Olson, P., 1986. The spectrum of sub-tidal variability in Chesapeake Bay circulation. *Estuarine, Coastal and Shelf Science*, **23** (4), 527–550.

Prandle, D., 1981. Salinity intrusion in estuaries. *Journal of Physical Oceanography*, **11**, 1311–1324.

Prandle, D., 1982. The vertical structure of tidal currents and other oscillatory flows. *Continental Shelf Research*, **1** (2), 191–207.

Prandle, D., 1985. On salinity regimes and the vertical structure of residual flows in narrow tidal estuaries. *Estuarine Coastal and Shelf Science*, **20**, 615–633.

Prandle, D., 1997. The dynamics of suspended sediments in tidal waters. *Journal of Coastal Research* (Special Issue No. 25), 75–86.

Prandle, D., 1998. Global expressions for seasonal temperatures of the sea surface and ambient air: the influence of tidal currents and water depth. *Oceanologica Acta*, **21** (3), 419–428.

Prandle, D., 2004a. Saline intrusion in partially mixed estuaries. *Estuarine, Coastal and Shelf Science*, **59**, 385–397.

Prandle, D., 2004b. How tides and river flow determine estuarine bathymetry. *Progress in Oceanography*, **61**, 1–26.

Prandle D. and Lane, A., 1995. The annual temperature cycle in shelf seas. *Continental Shelf Research*, **15** (6), 681–704.

Pritchard, D.W., 1955. Estuarine circulation patterns. *Proceedings of the American Society of Civil Engineers*, **81** (717), 1–11.

Rigter, B.P., 1973. Minimum length of salt intrusion in estuaries. *Proceedings of the American Society of Civil Engineers Journal of the Hydraulics Division*, **99**, (HY9), 1475–1496.

Rippeth, T.P., Fisher, N.R., and Simpson, J.H., 2001. The cycle of turbulent dissipation in the presence of tidal straining. *Journal of Physical Oceanography*, **31**, 2458–2471.

Rossiter, J.R., 1954. The North Sea Storm Surge of 31 January and 1 February 1953. *Philosophical Transactions of the Royal Society of London*, A, **246**, 317–400.

Schultz, E.A. and Simmons, H.B., 1957. Fresh water–salt water density currents, a major cause of siltation in estuaries. *Technical Bulletin*, No. 2, *Communication on Tidal Hydraulics*, U.S. Army, Corps of Engineers.

Simpson, J.H. and Bowers, D.G., 1981. Models of stratification and frontal movement in shelf seas. *Deep-Sea Research*, **28**, 727–738.

Simpson, J.H. and Hunter, J.R., 1974. Fronts in the Irish Sea. *Nature*, **250**, 404–406.

Simpson, J.H., Brown, J., Matthews, J., and Allen, G., 1990. Tidal straining, density currents and stirring in the control of estuarine stratification. *Estuaries*, **13** (2), 125–132.

Souza, A.J. and Simpson, J.H., 1997. Controls on stratification in the Rhine ROFI system. *Journal of Marine Systems*, **12**, 311–323.

Uncles, R.J., Stephens, J.A. and Smith, R.E., 2002. The dependence of estuarine turbidity on tidal intrusion length, tidal range and residence time. *Continental Shelf Research*, **22**, 1835–1856.

Wang, D.P. and Elliott, A.J., 1978. Non-tidal variability in Chesapeake Bay and Potomac River: evidence for non-local forcing. *Journal of Physical Oceanography*, **8** (2), 225–232.

# 5

# Sediment regimes

## 5.1 Introduction

Understanding and predicting concentrations of suspended particulate matter (SPM) in estuaries are important because of their impact on (i) light occlusion and thereby primary production, (ii) pathways for adsorbed contaminants and (iii) rates of accretion and erosion and associated bathymetric evolution.

Sediments are traditionally classified according to their (mass equivalent) particle diameter, $d$, with sizes ranging from clay <4 µ, 4 µ < silt < 63 µ, 63 µ < sand < 1000 µ through to gravel and rocks. Importantly, the silt–sand demarcation separates cohesive (mutual attraction by electro-chemical forces) from non-cohesive sediments. In higher concentrations, the former tend to flocculate into multiple assemblages which can both settle more rapidly and inhibit the upward flux of turbulent energy from the sea bed (Krone, 1962). In extreme cases, a layer of liquid mud may form a two-phase flow continuum. Moreover, once deposited, consolidation of cohesive material can radically change re-erosion rates. Only a few percent of 'mud' content may strongly influence what might appear to be a cohesionless sandy bed (Winterwerp and van Kesteren, 2004).

Over millennia, the inter-glacial rise and fall of sea level effectively determines estuarine morphology. Over shorter time scales, of interest to coastal engineers and coastal planners, some quasi-equilibria develop encompassing variations in bathymetry over ebb to flood and spring to neap tides, alongside seasonal cycles, random storms and episodic extreme events. For efficient management of estuaries, we need to understand the associated patterns of sediment movement in order to harmonise development with natural trends and to mitigate related hazards such as flooding or silting of navigation channels.

### 5.1.1 Sediment dynamics

At first sight, sediment dynamics appear deceptively simple – a sequence of erosion, suspension, transport and deposition. However, severe complications can arise when encompassing a mixture of sediment sizes influenced by past and present dynamics, modulated by chemical and biological processes. The limits of observational technology in measuring parameters such as concentrations, size-spectra, fluxes or net scour/accretion exacerbate these complications.

The predominant influences on sediment regimes in estuaries are tidal and storm currents, enhanced in exposed shallow water by wave stirring. Detailed accounts of the mechanics of sediment motion associated with tidal currents and waves can be found in Grant and Madsen (1979), Van Rijn (1993) and Soulsby (1997). Postma (1967) describes general features of the erosion, deposition and intervening transport of SPM in tidal regimes. For all but the coarsest grain sediment, several cycles of ebb and flood movement may occur between erosion and subsequent deposition. Hence, deposition can occur over a wide region beyond the source. Since time in suspension increases for finer, slowly settling material, such mechanisms may contribute to a residue of fine materials on tidal flats and to trapping of coarser material in deeper channels.

### 5.1.2 *Modelling*

Reproducing these characteristics in models remains sensitive to largely empirical formulae used in prescribing erosion and deposition rates. Bed roughness strongly influences these rates; it is largely determined by the composition (fine to coarse) and form (ripples, waves) of the bed. Sediment processes are complicated by the continuous dynamical feedback between this roughness and the overlying vertical structure of tidal currents and waves and their associated turbulence regimes (Fig. 5.1). Bed roughness can change significantly over both the ebb to flood and the neap to spring tidal cycles. Associated erosion and deposition rates may then

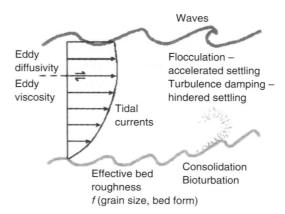

Fig. 5.1. Processes determining sediment erosion, transport and deposition.

vary considerably over these cycles and dramatically over seasons or in the course of a major event.

Conventionally, erosion is assumed to occur when the bed shear forces exceed the resistance of the bed sediment, characterised by a 'critical shear stress for erosion'. The rate of erosion is generally assumed to be proportional to the excess of the applied shear stress over the critical shear stress (Partheniades, 1965). Associated formulae vary widely, with consequent erosion rates sometimes varying by factors of 10 or more, emphasising the difficulty in formulating robust portable models.

In nature, this threshold depends on particle size distribution and both chemical and biological modulation, including effects of bioturbation and biological binding. Bioturbation of the top metre or so of surficial sediment may significantly reduce erosion thresholds. Conversely, (surficial) biological binding can have the opposite effect – especially in inter-tidal zones (Romano *et al.*, 2003). Erosion depends not only on the prevailing physical, chemical and biological composition but also on the conditions at the time of deposition and the intervening historical sequence.

Subsequent settlement of particles depends on their size, density and the ambient regime of turbulence and chemical forces in the surrounding water. Sedimentation is usually assumed to occur when quiescent dynamical conditions are below some threshold for erosion at a rate equal to the product of the near-bed concentration and the 'settling velocity'. The latter can be estimated from the density, size and shape of the sediment or determined in laboratory settling tubes.

Accurate simulation of sediment fluxes requires an initial prescription of the distribution of surficial sediments. Simulations over larger and longer space and time scales need to incorporate sequential changes in these surficial sediments as these adjust to variations in tidal and wave conditions resulting from trends and cycles in the inter-related evolving bathymetry. On even longer time scales, likely changes in both msl and sources of marine and fluvial sediments need to be incorporated.

### *5.1.3 Approach*

Here we focus on tractable elements, namely regimes dominated by tidal forcing. Concentrating on tidal components provides inter-comparisons between results from theory, models and observations in terms of robustly determined constituents. The analytical solutions obtained are intended to provide insight into mechanisms and thereby complement rather than substitute for existing complex numerical models.

SPM concentration models involve the solution of the 'conservation of mass' equation:

$$\frac{dC}{dt} = \frac{\partial C}{\partial t} + U\frac{\partial C}{\partial X} + V\frac{\partial C}{\partial Y} + W\frac{\partial C}{\partial Z} = \frac{\partial}{\partial Z}K_z\frac{\partial C}{\partial Z} - \text{sinks} + \text{sources}, \quad (5.1)$$

where $C$ is concentration, $U$, $V$ and $W$ are velocities along orthogonal axes $X$, $Y$ and $Z$ (vertical) and $K_z$ is a vertical eddy dispersion coefficient.

The horizontal advective velocities $U$ and $V$ can be accurately calculated in tidal models, as described in Chapters 2 and 3. The omission of axial and transverse dispersion terms in (5.1) is generally valid in models with sufficiently fine resolution in time and space.

Here we assume that erosion is proportional to some power of velocity and no thresholds are introduced for either erosion or deposition. Moreover, erosion and deposition are allowed to occur simultaneously. Analytical solutions of (5.1) show how suspension is determined by three parameters, namely the settling or fall velocity $W_s$, vertical turbulent displacements characterised by a vertical eddy dispersion coefficient $K_z$ and water depth $D$. These solutions provide Theoretical Frameworks illustrating the nature and scaling parameters that determine SPM concentrations, times in suspension, vertical profiles and the predominant tidal constituents in time series of SPM.

In strongly tidal conditions without pronounced stratification, the magnitude of the vertical diffusivity may be approximated by that of the vertical eddy viscosity coefficient. For consistency of notation with the original papers by Prandle (1997a and 1997b), henceforth we replace $K_z$ by $E$. In order to derive simplified analytical solutions, it is assumed that $E$ is constant both temporally and vertically.

The analytical expressions for erosion and deposition of SPM are presented in Sections 5.2 and 5.3. The mathematical details involved in combining these into expressions for resulting suspended concentrations are shown in Appendix 5A and outlined in Section 5.4. Section 5.5 describes the effects of integrating continuous cycles of erosion, modulated by an exponential settling rate to determine characteristic tidal spectra of SPM concentrations. Applications of these theories are examined in Section 5.6 against both modelled and observed results. Section 5.7 reflects on related progress towards predicting bathymetric evolution in estuaries.

## 5.2 Erosion

Erosion $ER(t)$ due to a tidal current $U(t)$ is generally assumed to be of the form

$$ER(t) = \gamma \rho f |U(t)|^N; \qquad \text{for } |U(t)| > U_c \qquad (5.2)$$

$$ER(t) = 0; \qquad \text{for } |U(t)| \leq U_c, \qquad (5.3)$$

where $\gamma$ is an empirical coefficient, $\rho$ the density of water, $f$ the bed friction coefficient and $N$ is some power of velocity typically in the range of 2–5.

The stipulation of a minimum threshold velocity, $U_c$, below which no erosion occurs, often has surprisingly little effect in strong tidal conditions where velocities exceed $0.5\,\mathrm{m\,s^{-1}}$. As an example, setting $U_c = 0.5U^*$ for a tidal velocity $U(t) = U^* \cos \omega t$ reduces the net erosion by only 10% for $N = 4$. Thus, for simplicity, it is convenient henceforth to set $U_c = 0$. Lavelle *et al.* (1984) carried out similar analyses of (5.2) and (5.3) and inferred a value of $N = 8$ for fine sediment; they also omitted $U_c$. Van Rijn (1993) found transport rates of fine sands proportional to powers of $U$ between 2.5 and 4, dependent on wave conditions.

Lane and Prandle (2006) found, for $N = 2$, a value of $\gamma = 0.0001$ $(\mathrm{m^{-1}\,s})$ best reproduced sediment concentrations in the Mersey Estuary.

### 5.2.2 Spring–neap cycle

A characteristic of tidal currents in mid-latitudes is the predominance of the semi-diurnal lunar $M_2$ and solar $S_2$ constituents with periods 12.42 h and 12 h, respectively (see Appendix 1A). Although the ratio of their tidal potentials is $1:0.46$, their observed ratio at the coast is generally closer to $1:0.33$ (related to the proportionately greater frictional dissipation effective for all constituents other than $M_2$). The small difference in their periods produces the widely observed 15-day spring (in phase) and neap (out of phase) $MS_f$ tidal constituent alongside the related quarter-diurnal constituent $MS_4$.

For the case of $N = 2$ and $U_c = 0$, the erosional time series for a combination of $M_2$ and $S_2$ tidal currents of amplitude $U_M$ and $U_s$, alongside a constant residual velocity $U_0$, is

$$
\begin{aligned}
(U_M \cos M_2 t + U_S \cos S_2 t + U_0)^2 &= 0.5 U_M{}^2 (1 + \cos M_4 t) \\
&\quad + 0.5 U_S{}^2 (1 + \cos S_4 t) + U_0{}^2 \\
&\quad + 2U_0 (U_M \cos M_2 t + U_S \cos S_2 t) \\
&\quad + U_M U_S (\cos MS_f t + \cos MS_4 t). \quad (5.4)
\end{aligned}
$$

The frequencies $M_4 = 2M_2$, $S_4 = 2S_2$ and $MS_4 = M_2 + S_2$ represent quarter-diurnal constituents while $MS_f = S_2 - M_2$ has a period of 15 days.

Table 5.1 lists the tidal constituents corresponding to (5.4). To illustrate typical relative magnitudes of these constituents, it is assumed that $U_M = 3U_S = 1$ (arbitrary units). Further assuming $U_0 \ll U_M$, the largest erosional constituents are $Z_0$, $0.55\,M_4$, 0.5 and both $MS_4$ and $MS_f$, 0.33.

Thus, while tidal currents predominantly involve $M_2$ and $S_2$ constituents, these translate into time series of erosion characterised in descending order by $Z_0$, $M_4$, $MS_4$ and $MS_f$. However, this erosional time series is subsequently modulated by the deposition phase – resulting in further modulation of the SPM concentrations as described in Section 5.5.

The ratio of the erosional amplitudes of $M_4 : MS_4$ is $U_M : 2U_S$; thus, the relative phasing of these constituents over the spring–neap cycle can strongly influence the apparent characteristics of the time series. The two constituents are in phase at

Table 5.1 *Amplitudes of erosional constituents*
*corresponding to (5.4) for* N = 2 *and* $U_M = 3 U_S = 1$

| N=2 | | $U_M = 3U_S = 1$ |
|---|---|---|
| $Z_0$ | $0.5(U_M^2 + U_S^2) + U_0^2$ | $0.55 + U_0^2$ |
| $MS_f$ | $U_M U_S$ | 0.33 |
| $M_2$ | $2U_M U_0$ | $2U_0$ |
| $S_2$ | $2U_S U_0$ | $0.67U_0$ |
| $M_4$ | $0.5U_M^2$ | 0.50 |
| $MS_4$ | $U_M U_S$ | 0.33 |
| $S_4$ | $0.5U_S^2$ | 0.05 |

spring tides – indicating a strong quarter-diurnal variability. Conversely, they are out
of phase at neap tides, often resulting in a predominant semi-diurnal constituent that
can be incorrectly interpreted as suggesting either predominant diurnal currents or
horizontal advective influences.

### *Influence of* $U_0$

From Table 5.1, the two largest constituents associated with $U_0$ will be $M_2$ and $S_2$.
For these constituents to be equal to the $M_4$ constituent requires $U_0$ to be 0.25 of the
$M_2$ current amplitude or 0.75 of the $S_2$ amplitude.

### 5.2.3 Erosion in shallow cross sections

Section 2.6.3 shows that to maintain continuity of an $M_2$ flux $U^* \cos(\omega t)$ through a
triangular cross section of mean depth $D$ and elevation, $\varsigma^* \cos(\omega t - \theta)$, requires
currents at $M_4$ and $Z_0$ frequencies given by

$$U_2 = -a\,U^* \cos(2\omega t - \theta) \qquad \text{and}$$
$$U_0 = -a\,U^* \cos\theta, \tag{5.5}$$

where $a = \varsigma^*/D$.

Assuming erosion proportional to velocity squared, this yields erosional components at the following frequencies:

$$[U^* \cos(\omega t) - a\,U^* \cos(2\omega t - \theta) - a\,U^* \cos\theta]^2 =$$
$$U^{*2}[(0.5(1 + a^2) + a^2 \cos^2\theta)$$
$$- a\,(\cos(\omega t - \theta) - 2\cos\theta\cos\omega t)$$
$$+ 0.5\cos(2\omega t) + 2a^2 \cos\theta \cos(2\omega t - \theta)$$
$$- a\cos(3\omega t - \theta)$$
$$+ 0.5a^2 \cos(4\omega t - 2\theta)]. \tag{5.6}$$

Since the value of '*a*' can approach 1 in shallow, strongly tidal estuaries, significant current components can occur at all of the above frequencies, i.e. $Z_0$, $M_2$, $M_4$, $M_6$ and $M_8$.

### 5.2.4 Advective component

At any fixed position, advection may constitute a source or sink of sediments. The differing tidal characteristics of advection in comparison with localised resuspension are described below.

For a purely advective source, (5.1) reduces to

$$\frac{\mathrm{d}C}{\mathrm{d}t} = -U\frac{\partial C}{\partial X} \tag{5.7}$$

with the solution

$$C = -\frac{U^*}{\omega}\sin\omega t \quad \frac{\partial C}{\partial X} \tag{5.8}$$

for the current component $U^*\cos\omega t$ and a constant horizontal gradient $\partial C/\partial X$.

Significant values of $\partial C/\partial X$ are likely where spatial inhomogeneity in tidal currents, sediment supply or water depths occur. The resulting tidal constituents for suspended concentrations are proportional to the product of tidal current amplitude and period (of the constituent concerned) and show a 90° phase shift relative to the current.

## 5.3  Deposition

### 5.3.1  Advective settlement versus turbulent suspension (Peclet number)

Siltation or deposition occurs both by steady advective settlement at the fall velocity $W_S$ and by intermittent contacts with the bed via vertical 'turbulent excursions' characterised by the vertical dispersion coefficient, $K_z$, here represented by $E$. Specific near-bed conditions determine entrainment or 'capture' rates.

The relative importance of advection via the fall velocity $W_S$, compared with dispersion via $E$, may be estimated from their associated time constants. The time $T_A$ to fall from the surface to the bed via advection is $D/W_S$ while the time $T_E$ to mix vertically following erosion at the bed is $D^2/E$ (Lane and Prandle, 1996). Thus, the ratio of the time scales for advective settling to turbulent suspension is $E/DW_S$, i.e. the Peclet number for vertical mixing. The Peclet number is inversely proportional to the familiar Rouse number. In shallow, strongly tidal waters with an absence of pronounced stratification, a spatially and temporally constant approximation to the vertical eddy viscosity and eddy dispersion coefficients can conveniently be represented by (Prandle, 1982):

$$K_z = E = fU^* D, \qquad (5.9)$$

where $f$, the bed friction coefficient, is approximately 0.0025 and $U^*$ is the tidal current amplitude. The Peclet number is then $f U^*/W_S$.

Assuming axial and lateral homogeneity allows the terms involving horizontal advection in (5.1) to be neglected. A 'single-point' vertical distribution of suspended sediments can then be described by the dispersion equation:

$$\frac{\partial C}{\partial t} + W_s \frac{\partial C}{\partial Z} = E \frac{\partial^2 C}{\partial Z^2} - \text{sinks} + \text{sources}, \qquad (5.10)$$

where the sediment fall velocity $W_s$ replaces the (small) vertical velocity $W$.

### 5.3.2 Bottom boundary conditions

A major difficulty in numerical simulations of suspended sediments based on (5.10) is the representation of conditions close to the bed, where sharp gradients can exist for both $E$ and $W_s$ (flocculation) and hence in concentration. Appendix 5A shows that the assumption of settlement at the rate $W_s C_o$ is generally twice the value pertaining with a Gaussian vertical sediment distribution where dispersion counteracts half of this advective settlement.

The use here of an analytical solution avoids the sensitivity, encountered in numerical schemes, to the precise discretisation of the near-bed region. However, questions still arise as to the extent to which suspended particles rebound or settle on collision with the bed. Sanford and Halka (1993) provide a comprehensive analysis of alternative bottom boundary conditions.

Three conditions are examined in Appendix 5A:

[A] A fully reflective bed. Settlement occurs via advection at the rate $0.5W_S C_o$. Dispersive collisions with the bed rebound.
[B] A fully absorptive bed. Like condition [A], except that all dispersive collisions settle.
[C] Like [A], except that there is additional settlement, by dispersion, of those particles which have previously been reflected from the surface. This solution is an expedient approximation, providing solutions intermediate between [A] and [B].

The effect of these three boundary conditions on the rate of deposition is illustrated in Fig. 5.2 (Prandle, 1997a).

For $E/DW_S < 1$, Fig. 5.2 and Appendix 5A show that the deposition rate, of $0.5W_S C_0$, is the same for all three conditions.

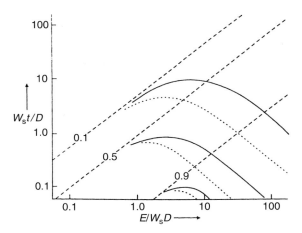

Fig. 5.2. Sediment fraction remaining in suspension at time $W_S\, t/D$ as a function of $E/W_S\, D$. Bottom boundary condition [A] is dashed, [B] is dotted and [C] is solid.

For $E/DW_S > 1$, the deposition rate varies widely for the three conditions. For conditions **[B]** and **[C]**, maximum concentrations occur in the range $1 < E/W_S D < 10$. Here, we adopt the intermediate approximation, bottom boundary condition **[C]**.

## 5.4 Suspended concentrations

### 5.4.1 Time series of sediment concentration profiles

Appendix 5A shows that, by adopting the boundary condition **[C]**, time series of sediment concentration profiles associated with each erosional event of magnitude M take the form

$$C(Z,t) = \frac{M}{(4\pi\, E\, t)^{1/2}} \left[ \exp -\frac{(Z + W_s\, t)^2}{4E\, t} + \exp -\frac{(2D + W_s\, t - Z)^2}{4E\, t} \right]. \quad (5.11)$$

The observed time series represents a time integration of all such preceding 'events' as described in Section 5.5.

In developing generalised theory, it is advantageous to derive expressions for deposition in terms of depth–mean concentrations. From (5A.4) and (5A.7),

$$\frac{C_{z=0}}{\bar{C}} \approx \frac{D}{\sqrt{\pi E t}}\ \frac{\exp\left(-W_s^2 t^2 / 4Et\right)}{\left[1 - \left(-W_s^2 t^2 / 4Et\right)^{1/2}\right]} \approx \frac{D}{\sqrt{\pi E t}}. \quad (5.12)$$

Thus, a deposition rate expressed as $W_s C_{z=0}$ transforms to $0.56 W_s (D/E)^{1/2} (\bar{C}/t^{1/2})$ when related to depth-averaged concentrations.

### 5.4.2 Exponential deposition, half-lives in suspension, t$_{50}$

Approximating deposition of an initial suspended concentration $C_0$ by an exponential loss rate $C_0 e^{-\alpha t}$ enables simple analytical expressions to be derived for the time series of combined erosion and deposition over successive tidal cycles. Such an exponential decay rate corresponds to a half-life in suspension $t_{50} = 0.693/\alpha$.

For $E/W_s D < 1$.

It is shown in Appendix 5A, (5A.7) and from Fig. 5.2, that the fraction of sediment remaining in suspension $FR$ approximates

$$FR = 1 - 0.5 \left( \frac{W_s^2 t}{E} \right)^{1/2}. \tag{5.13}$$

Equating (5.13) to $e^{-\alpha t}$ for $FR = 0.5$ requires

$$\alpha = 0.693 \frac{W_s^2}{E}. \tag{5.14}$$

For $E/W_s D > 1$

Adopting condition [C], Appendix 5A indicates an expression for the half-life $t_{50} = 0.693/\alpha$, i.e. the time required for 50% deposition is equivalent to

$$\alpha = \frac{0.1 E}{D^2}. \tag{5.15}$$

For implementation of (5.14) and (5.15), a continuous transition can be obtained by simple curve fitting of the results shown in Fig. 5.2. Thus, the parameter $\alpha$ can be closely approximated by

$$\alpha = \frac{0.693 \, W_s/D}{10^x}, \tag{5.16}$$

where $x$ is the root of the equation:

$$x^2 - 0.79 \, x + j(0.79 - j) - 0.144 = 0 \tag{5.17}$$

with $j = \log_{10} E/W_s D$.

Assuming the following relationship between fall velocity $W_S$ and particle diameter $d$

$$W_s \, (\text{ms}^{-1}) = 10^{-6} \, d^2 \, (\mu\text{m}). \tag{5.18}$$

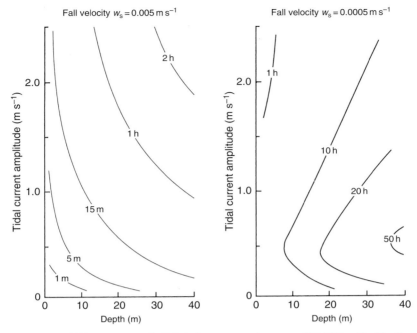

Fig. 5.3. Half-lives, $t_{50}$, of SPM in suspension as $f(D, U^*)$. (Left) Sand, $W_S = 0.005\,\mathrm{m\,s}^{-1}$, (right) silt, $W_S = 0.0005\,\mathrm{m\,s}^{-1}$. $t_{50} = 0.693/\alpha$, $\alpha$ from (5.16).

Figure 5.3 illustrates corresponding half-lives in suspension for $W_S = 0.005$ and $0.0005\,\mathrm{m\,s}^{-1}$, i.e. $d = 71$ and $22\,\mu$, as functions of tidal current amplitude and depth. Thus, for sand, half-lives range from a minute in sluggish shallow water to $2\,\mathrm{h}$ in deep strongly tidal waters. For fine silt, these values range from hours in shallow water to 2 days in deep water.

More generally, using dynamical solutions for a 'synchronous' estuary, Fig. 7.6 indicates values of $\alpha$ ranging for sand from $10^{-3}$ to $10^{-2}\,\mathrm{s}^{-1}$ corresponding to values for $t_{50}$ of 10–1 m. The equivalent values for fine silt and clay are $\alpha$ from $10^{-6}$ to $10^{-4}\,\mathrm{s}^{-1}$ for $t_{50}$ of 200–2 h.

An important result is that for the finer sediment, the half-life in suspension is almost always of order $(6\,\mathrm{h})$ or greater. This implies that where estuarine sediment regimes are maintained by the influx of fine marine sediments, these will remain in near-continuous suspension and so their distributions will share features of a continuously suspended conservative tracer such as salt.

To further illustrate the characteristics and significance of this parameter $E/W_S D$, we use (2.19) to relate tidal current amplitudes $U^*$ to elevation amplitude $\zeta^*$. Figure 5.4 (Prandle, 2004) then shows, for $D = 8\,\mathrm{m}$, the relationship between $W_S$ and $E/W_S D$ for $\zeta^* = 1$, 2 and 3 m, i.e. almost the complete range of tidal conditions encountered.

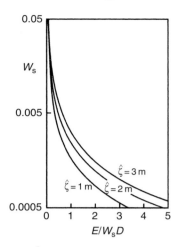

Fig. 5.4. Fall velocity $W_S$ (m s$^{-1}$) as a function of $E/W_s D$ tidal elevation amplitudes $\zeta^* = 1$, 2 and 3 m; results are for depth $D = 8$ m.

This illustrates how the demarcation line of $E/W_S D \approx 1$ coincides with fall velocities of order of $(1 \text{ mm s}^{-1})$ or, from (5.18), $d = 32\,\mu$. It also shows how the value of $E/W_S D$ lies in the range from 0.1 to 10 for sediment diameters in the range 20–200 $\mu$. Hence, the approximation, (5.16), for $\alpha$, based on Fig. 5.2, should be widely applicable.

### 5.4.3  Vertical profile of suspended sediments

It is shown in Appendix 5A that for $E/DW_S < 0.3$, less than 1% of particles reach the sea surface, whereas for $E/DW_S > 10$, suspended sediments are well mixed vertically. Thus, throughout the range of conditions considered, clay is always likely to be well mixed vertically, whereas sand will be confined to near-bed regions for tidal velocity amplitudes $U^* \leq 0.5 \text{ m s}^{-1}$. Likewise, pronounced vertical structure throughout the water column is likely for silt when $U^* \leq 0.5 \text{ m s}^{-1}$.

A continuous functional description of the vertical profiles of suspended sediment concentrations was calculated by Prandle (2004) by numerical fitting of a profile e$^{-\beta z}$ (where $z$ is the fractional height above the bed) to simulations based on (5.12). The following expression for $\beta$ was derived:

$$\beta = \left[ 0.91 \, \log_{10} \left( 6.3 \, \frac{E}{DW_s} \right) \right]^{-1.7} - 1. \tag{5.19}$$

Figure 5.5(a) (Prandle, 2004) shows values of $\beta$ over the range $E/DW_S$ from 0 to 2. Figure 5.5(b) shows corresponding values of sediment profiles e$^{-\beta z}$ $(1 - e^{-\beta})/\beta$,

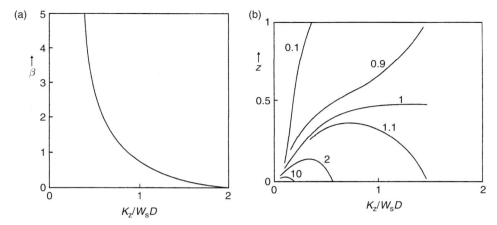

Fig. 5.5. Concentration profiles and values of $\beta$ described by sediment profile $e^{-\beta z}$, (5.19). (a) Values of $\beta$ as a function of $K_z/W_sD$. (b) Corresponding profiles; contours are fractions of depth-mean concentrations, surface is $z = 1$.

illustrating how effective complete vertical mixing is achieved for $E/DW_S > 2$, and 'bed load' only occurs for $E/DW_S < 0.1$. (Concentration profiles are normalised relative to the depth-averaged value.)

## 5.5  SPM time series for continuous tidal cycles

The concentration $C(t)$ associated with erosion varying sinusoidally at a rate of $\cos \omega t$ subject to an exponential decay rate $-\alpha C$ involves integration over all preceding time $t'$ from $-\infty$ to $t$, that is

$$C(t) = \int_{-\infty}^{t} \cos \omega t' e^{-\alpha(t-t')} dt' = \frac{\alpha \cos \omega t + \omega \sin \omega t}{\alpha^2 + \omega^2}. \qquad (5.20)$$

Hence, from (5.2), the concentration, $C_\omega$, for any erosional constituent, $\omega$, is given by

$$C_\omega = \frac{\gamma f \rho \, [U^N]_\omega}{D(\omega^2 + \alpha^2)^{1/2}}, \qquad (5.21)$$

where $[U^N]_\omega$ is the erosional amplitude at frequency $\omega$ as shown in Table 5.1 for $N = 2$.

Thus, erosion generated at each of the tidal constituents in the expansion of source terms is subsequently modulated by an exponential decay rate that involves an amplitude reduction by the factor $(\alpha^2 + \omega^2)^{-1/2}$ and a phase-lag of $\arctan(\omega/\alpha)$.

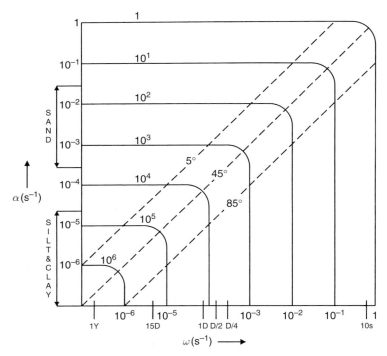

Fig. 5.6. Concentration amplitudes for unit erosion at cyclic frequencies $\omega$ for deposition rate $e^{-\alpha t}$. Solid contours indicate relative amplitudes and dashed lines phase lag between SPM and erosion (5.20).

The response function represented by (5.20) is shown in Fig. 5.6 (Prandle, 1997b) for a range of values of $\alpha$ and $\omega$ from $10^{-7}$ to $1\,\text{s}^{-1}$. For $\alpha \gg \omega$, the amplitude of the response is proportional only to $1/\alpha$ with zero phase lag. Conversely, for $\omega \gg \alpha$, the amplitude of the response is proportional only to the period concerned with a $90°$ phase lag.

Thus, for equal amounts of eroded sand ($\alpha \sim 10^{-3}\,\text{s}^{-1}$) and silt/clay ($\alpha \sim 10^{-6}\,\text{s}^{-1}$), away from the immediate vicinity of the bed, the amplitude ratios of the suspended sediment tidal signal for sand are always much smaller than for silt/clay.

Within this silt/clay fraction, the ratio of the SPM tidal constituent amplitudes is determined by the product of (i) the ratios shown in Table 5.1 and (ii) the duration of the tidal periods concerned. These amplitudes are largely independent of the specific value of $\alpha$ (within the range concerned). Likewise, for this silt/clay fraction, the phases are lagged by $90°$ in the quater-diurnal to diurnal tidal band. Thus, factoring the amplitudes in Table 5.1 by tidal period, we expect the $MS_f$, 15 days cycle to predominate followed by $M_4$ and $MS_4$. Phase values of up to $90°$ for $MS_f$ indicate associated maximum suspended sediment concentrations occurring up to 3.5 days after maximum currents. Conversely, phase values approaching $0°$ for $M_2$ and $S_2$ indicate associated maximum suspensions in phase with maximum currents.

For $\omega \gg \alpha$, the $Z_0$ constituent is factored by $\omega/\alpha$ relative to the tidal frequency. Thus, by using the $Z_0/MS_f$ ratio derived from observations alongside the expansions shown in Table 5.1 and (5.20), a direct estimate of $\alpha$ can be obtained.

## 5.6 Observed and modelled SPM time series

### 5.6.1 Observational technologies

Observations are crucial to developing and assessing SPM models. *In situ* concentrations are routinely monitored acoustically, optically and mechanically. Acoustic backscatter (ABS) probes provide vertical profiles of concentration, multi-frequency probes provide information on grain size – usually at a single point. Pumped samples, bottles and traps are used in mechanical devices. Recent developments of *in situ* laser particle sizers provide valuable information on particle spectra non-invasively (mechanical samplers can corrupt these spectra).

Available observations suffer from fundamental shortcomings, namely (i) calibration from sensor units to concentration involving complex sensitivity to particle size spectra in optical and acoustic instruments and to atmospheric corrections and sun angle effects in remote sensing; (ii) unresolved particle-size spectra and (iii) limited spatial and temporal coverage relative to the inhomogeneity of sediment distributions.

The spatial resolution of *in situ* concentration measurements is generally limited to single points (or limited profiles) in optical backscatter (OBS) and ABS sensors and to surface values from satellite or aircraft sensors. Techniques to circumvent these shortcomings are described by Gerritsen *et al.* (2000) where the spatial patterns of surface imagery are used to validate models.

Each instrument has its own calibration peculiarities. Moreover, all of these calibrations vary as the mean particle size changes. Optical devices rely on occlusion of light – transmittance or reflectance (OBS). Since this is dependent on the surface area of the particle, recordings are more sensitive to finer scale particles. Hence, observed concentrations need to be calibrated by reference to some representative particle radius. The plate-like character of flocs complicates such calibrations.

Conversely, ABS (in the range of frequencies used in ABS instruments) increases with particle volume, and hence, these instruments are more sensitive to coarse particles. The optical instruments also experience fouling, and all of the instruments can be swamped above certain concentrations.

Satellite images of (surface) SPM concentrations can be used in conjunction with model simulations to infer the magnitude of discrete sediment sources. Aircraft surveillance using multi-wavelength imagery can differentiate between the reflectance from SPM associated with chlorophyll and that from various sediment fractions. However, the need for atmospheric corrections introduces some reliance on *in situ* calibrations.

On the longer time scale, information in sediment cores (judicious choice of location is crucial) may be dated using seasonal striations, specific contaminants (radio nuclides, Pb-210, etc.) and various natural chemical signals or biological fossils (Hutchinson and Prandle, 1994). The range of such techniques is expanding rapidly providing opportunities to derive both geographic provenance and associated age. Both light detection and ranging (LIDAR) and synthetic aperture radar (SAR) surveys can be used to determine sequences of bathymetric evolution.

### 5.6.2  Observed time series

Figure 5.7 (Prandle, 1997b) shows three examples of simultaneous time-series recordings of suspended sediment and tidal velocity. Table 5.2 lists results from tidal analyses of

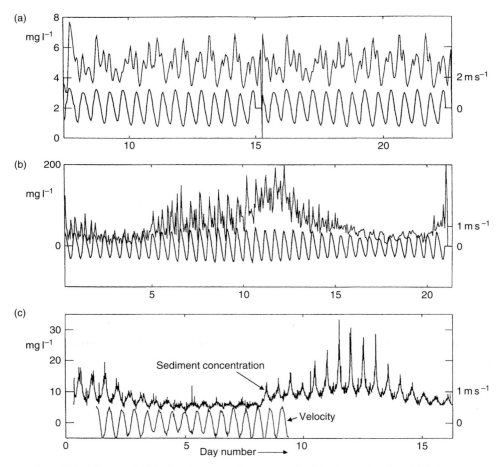

Fig. 5.7. Observed SPM and current time series in (a) Dover Straits, (b) Mersey Estuary and (c) Holderness Coast.

Table 5.2 *Tidal constituents of observed tidal currents and SPM (Fig. 5.7), currents in m s$^{-1}$, SPM in mg l$^{-1}$. Phases adjusted to 0$^{\circ}$ for M$_2$ and S$_2$*

| Constant | Mean | Amplitude (m s$^{-1}$); (mg l$^{-1}$) | | | | | Phase (degrees) | | | | |
|---|---|---|---|---|---|---|---|---|---|---|---|
| | | MS$_f$ | M$_2$ | S$_2$ | M$_4$ | MS$_4$ | MS$_f$ | M$_2$ | S$_2$ | M$_4$ | MS$_4$ |
| *Dover* | | | | | | | | | | | |
| U* | −8.1 | 10.2 | 92.2 | 24.7 | 10.8 | 3.2 | 90 | 0 | 0 | 76 | 81 |
| SPM | 4.21 | 0.98 | 0.25 | 0.27 | 0.39 | 0.23 | 41 | 96 | 158 | 352 | 354 |
| % | 100 | 23 | 6 | 6 | 9 | 5 | | | | | |
| *Holderness* | | | | | | | | | | | |
| U* | −0.6 | 1.2 | 49.6 | 17.6 | 0.9 | 1.5 | 62 | 0 | 0 | 357 | 341 |
| SPM | 8.75 | 3.39 | 2.03 | 1.26 | 0.69 | 0.78 | 16 | 44 | 70 | 118 | 121 |
| % | 100 | 39 | 23 | 14 | 8 | 9 | | | | | |
| *Mersey* | | | | | | | | | | | |
| U* | −0.8 | 1.8 | 60.6 | 19.9 | 11.5 | 7.0 | 32 | 0 | 0 | 234 | 274 |
| SPM | 61.5 | 40.6 | 12.9 | 6.3 | 13.4 | 10.7 | 19 | 28 | 25 | 20 | 11 |
| % | 100 | 66 | 21 | 10 | 22 | 17 | | | | | |

*Source:* Prandle, 1997b.

these observations. These examples were selected as illustrative of tidally dominated conditions and correspond to tranquil weather conditions. The Dover Strait is a highly (tidally) energetic zone, 30 km wide and up to 60 m deep, linking the North Sea to the English Channel with currents exceeding 1 m s$^{-1}$. The Mersey Estuary is a shallow (<20 m deep) estuary with tidal range up to 10 m; the measurements shown were taken in the narrow entrance channel, 1-km wide, 10-km long (Prandle *et al.*, 1990). The Holderness measurements were taken some 4 km offshore of a long, rapidly eroding coastline (glacial till). The Dover Strait and Mersey sediment recordings used transmissometers; the Holderness recordings were by an OBS.

In all three cases, the MS$_f$ constituent is largest – as anticipated from combining (5.4) and (5.21). Using the above theory to interpret both the current-SPM amplitude ratios and the phase lags for MS$_f$, Prandle (1997b) derived the following values of $\alpha$: Dover Strait −3 × 10$^{-6}$ s$^{-1}$, Holderness −2 × 10$^{-5}$ s$^{-1}$ and Mersey −2 × 10$^{-5}$ s$^{-1}$. Jones *et al.* (1994) show that the spectral peak in the sediment distribution in the Dover Strait corresponds to a settling velocity of 10$^{-4}$ m s$^{-1}$. The Mersey and Holderness are likely to contain more coarse grained components.

In all three cases, the M$_2$ or M$_4$ constituent is next largest as anticipated earlier. Likewise, the phase values for all constituents (relative to the associated current values) are generally in the range of 0–90°. However, precise correspondence between these observed results and the theory developed here is complicated by such factors as the spectral width of settling velocities of the particles involved, finite supply, advection and vertical variations in concentration.

The mean concentration in the Mersey is an order of magnitude higher than in the Dover Strait. Concentrations at Holderness lie between these two. Thus, there is a suggestion of limited supply in the Dover Strait; moreover, the phase relationship for the $M_2$ constituent is indicative of a significant advective component.

### 5.6.3 Modelled time series

Figure 5.8 (Prandle, 2004) shows characteristic model-generated spring–neap cycles of SPM associated with localised resuspension in a tidally dominated estuary (Prandle, 1997a) for $E = K_z = 0.1$ and 10 $W_SD$. For $E = 0.1$ $W_SD$, particles scarcely reach the surface and with a short half-life in suspension, the quarter-diurnal constituent predominates. Conversely for $E = 10$ $W_SD$, particles are evenly distributed through

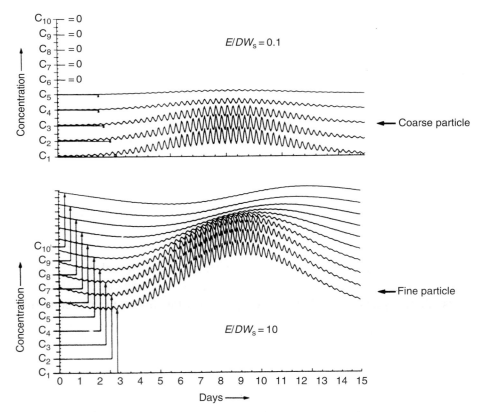

Fig. 5.8. Model simulations of SPM over a spring–neap tidal cycle. C1, C2, ... , C10 concentrations at fractional heights $z^{1/2} = 0.05$–$0.95$. (Top) $K_z/DW_S = 0.1$; (bottom) $K_z/DW_S = 10$.

the water column and the extended half-life in suspension amplifies the $MS_f$ constituent relative to $M_4$.

Fine resolution 3D estuarine models of tidal propagation can provide reliable and accurate simulations of water levels and flows, often limited only by inadequacies in the accuracy and resolution of the bathymetry. Likewise, recent developments in coupling these tidal dynamics with turbulence-closure modules provide good, detailed descriptions of current structure. However, fundamental uncertainties in modelling SPM fluxes arise from (i) insufficient information on surficial sediment distributions; (ii) the inherent complexities of erosion and deposition processes – especially regarding the influence of flocculation and (iii) the paucity of observational data for calibration/evaluation. Algorithms relating bed stresses with rates of erosion are plentiful but cannot easily embrace the full range and mixture of bed materials or the chemical and biological influences. Likewise, in conditions of high (near-bed) concentrations, sediment particles interact (flocculation and hindered settlement) modifying both the dynamics and the wider sediment regime.

The biggest difficulty in long-term simulations of SPM in estuaries is the specification of the available sources of sediments – internal, coastal, riverine and organic. Lack of information on particle size distributions of suspended sediments is a major deficiency along with ignorance of the nature and distribution for surficial sediments. The use of aircraft surveillance to provide SPM distributions (sea surface) may, via assimilation techniques, help to address this issue. While large area, long-term model simulations can link these sources, specific sequences of events need to be reproduced – resulting in practice in accumulation of errors.

The ultimate goal is to understand the evolving morphological equilibrium between bathymetry and the prevailing tidal currents, waves and mean sea level. Simulating net long-term bathymetric evolution is especially difficult, since it involves the net temporal integration of the spatial divergence of all simulations. The requirement for models to reproduce near-zero net accretion and erosion provides a demanding criterion. It is generally concluded that extrapolation using primitive equation (bottom-up) models is severely limited. Arguably, such models should only be used to indicate pathways and likelihoods of erosion/accretion for periods of a few years ahead. Further extrapolation often makes use of geomorphological (top-down) models which derive stability rules based on geological evidence of coastal movements. Bottom-up models can be used to test the validity of these top-down rules in any particular estuary.

The scope of monitoring, modelling and theoretical developments described highlights the need for collaboration interdisciplinary and internationally. The value of basic process studies in near-full scale flumes alongside extended estuarial observational campaigns is evident. Such data need to be quality assured and stored systematically for accessible dissemination. Permanent comprehensive observational networks exploiting synergistic aspects of a range of instruments/platforms integrally linked to modelling requirements/capabilities are especially valuable. The ultimate derivation of robust, portable SPM models depends on the availability of a range of such data sets from a wide range of estuaries, i.e. parallel programmes across estuaries with varying latitudes, scales, geological types and environmental exposures.

## 5.7  Summary of results and guidelines for application

Focusing on strongly tidal estuaries, by integrating erosion, suspension and deposition, analytical solutions are derived describing suspended sediment time series andtheir vertical structure. The related scaling parameters indicate the sensitivities to sediment type, tidal current speed and water depth, providing insight into and interpretation of sediment regimes obtained from observations or numerical simulations.

The leading question addressed is:

***How are the spectra of suspended sediments determined by estuarine dynamics?***

The major interest with sediment regimes in estuaries generally concerns bathymetric evolution. Additional ecological interests include (i) the transport of contaminants by adsorption onto fine sediments and (ii) light occlusion from high concentrations of suspended sediments. Geologists examine morphological evolution over millions of years and geomorphologists over millennia. Conversely, coastal engineers are concerned with relatively immediate impacts following extreme 'events' or 'interventions' and decadal adjustments to changes in msl and other long-term trends. Chapter 7 introduces a link between present-day morphological equilibria and historical changes in msl since the last ice age.

The earliest engineering studies of sediment transport were concerned with maintaining bathymetric equilibrium in the unidirectional flow regimes of irrigation and navigation canals. By contrast, one of the notable features in strongly tidal oscillatory regimes is the persistence over decades and re-emergence (after extreme events or 'intervention') of bathymetric features such as spits, pits and systematic inter-tidal forms.

Compared with tidal elevations, currents and salinity, suspended sediments exhibit extreme variability. This embraces (i) size, ranging from fine to coarse with particle diameters ranging from clay $< 4\,\mu$, silt $< 60\,\mu$, sand $< 1000\,\mu$ to gravels; (ii) temporal, spanning flood to ebb, neap to spring tides, from tranquil to storm conditions with related extreme values of water levels, velocities, waves and river flows; (iii) spatial, extending vertically, axially and transversally. Even without the technical difficulties of measuring concentrations of SPM, conditions just metres apart can show dramatic differences, reflecting the sensitivity to water depth, current speed, waves, turbulence intensity and surficial sediment distributions with their attendant bed features.

Recognising such intrinsic variability, the present approach is limited in scope, concentrating on first-order effects in strongly tidal regimes. This focus represents a tractable component of sediment dynamics, involving variability in concentrations on logarithmic scales and exploiting robustly determinable tidal characteristics. For a wider perspective, many additional features need to be considered, including the chemistry of fine particles (Partheniades, 1965), near-bed interactions between flow dynamics and sediments for coarse particles (Soulsby, 1997), wave–current interactions (Grant and Madsen, 1979) and the impacts of bed features (Van Rijn, 1993). Useful summaries are described by Davies and Thorne (2008) and Van Rijn (2007(a),(b) and (c)).

Postma (1967) emphasised the significance, in estuaries, of the separation between erosion at one location and the subsequent 'delayed' settlement at positions as far apart as the tidal excursion. Here, we aim to determine representative values of these delays, encapsulated by the 'half-life' of sediment in suspension, $t_{50}$. Rejecting the traditional demarcation into periods of erosion and deposition separated by thresholds of velocity or bed stress, 'single-point' analytical solutions for the simultaneous processes of erosion, suspension and deposition are derived. Similarly, the plethora of complicated formulae for erosion and settlement are simplified, with erosion related to some power of tidal velocity and deposition to the depth–mean concentration multiplied by an exponent based on the half-life in suspension.

By assuming that eddy diffusivity $K_z$ and eddy viscosity $E$ can be approximated by $K_z = E = fU^*D$, (3.23), these solutions indicate how the essential scaling of sediment motion is synthesised in the dimensionless parameter $E/W_S D$ as shown in Figs (5.4), (5.5) and (5.8), (where $f$ is the bed friction coefficient, $D$ is water depth, $U^*$ tidal current amplitude and $W_S$ sediment fall velocity). Turbulent diffusion, parameterised by the coefficient $E$, promotes the suspension of particles by random vertical oscillations, whereas the fall velocity $W_S$ represents steady advective settlement. The time taken for a particle to mix vertically by dispersion is $D^2/E$ whereas settlement by vertical advection occurs within $D/W_S$. Thus, the ratio of $E : W_S D$ reflects the relative times of deposition by advective settlement to that by diffusive vertical excursions.

### 5.7.1 Sediment suspension

For $E < 0.1 W_S D$, particles are confined to the near-bed region, whereas for $E > 10W_S D$, particles are evenly distributed throughout the water column (Fig. 5.8). Approximating settling velocities by sand, $W_S = 10^{-2}\,\mathrm{m\,s^{-1}}$; silt, $W_S = 10^{-4}\,\mathrm{m\,s^{-1}}$; and clay, $W_S = 10^{-6}\,\mathrm{m\,s^{-1}}$, it is shown that generally, sand is concentrated near the bed, clay is well mixed vertically and silt shows significant vertical structure. Assuming a sediment concentration profile $e^{-\beta z}$, an expression, (5.19), for $\beta$ in terms of $E/W_S\,D$ is derived. Corresponding vertical profiles are shown in Fig. 5.5.

### 5.7.2  Deposition

The rate of deposition is expressed by the function $Ce^{-\alpha t}$, where $\alpha = 0.693/t_{50}$ represents the exponential settling rate.

For $E \ll W_S\,D$, deposition is by advective settlement at a rate $(1/2)W_S C_0$ and $\alpha \sim 0.7\ W_S^2/E$, (5.14). The fractional rate of deposition is determined by $(1/2)$ $[(W_S^2 t)/E]^{1/2}$, i.e. 10% after $0.04E/W_S^2$, 50% after $E/W_S^2$ and 90% after $3.2E/W_S^2$ (Appendix 5A).

For $E \gg W_S\,D$, deposition is independent of $W_S$ but dependent on both the magnitude of the vertical dispersion coefficient and the precise near-bed conditions. Using an expedient bottom boundary condition, the value $\alpha \sim 0.1E/D^2$ is derived (5.15).

As $E \rightarrow W_S\,D$, the mean time in suspension approaches a maximum and hence both mean concentration and net transport will increase. Section 7.3.1 shows that this condition occurs for $W_S \sim 1\ \text{mm s}^{-1}$, i.e. particle diameter $d \sim 30\ \mu$.

Equation (5.17) provides a continuous expression for $\alpha$ across the full range of $E/W_S D$.

### 5.7.3  Tidal spectra of SPM

By integrating the above analytical solutions for SPM concentrations over successive tidal cycles, the spectrum of suspended sediments is calculated, (5.20). The characteristics of this spectrum are determined by the ratio of the exponential settlement rate, $\alpha$, to the frequency of the erosional tidal currents, $\omega$ (Fig. 5.6). For $\alpha > 10\omega$, the suspended sediment tidal amplitude is proportional to $1/\alpha$ with zero phase lag of SPM relative to current. Whereas, for $\omega > 10\alpha$, the amplitude is proportional to tidal period $(1/\omega)$ with a 90° phase lag.

From (5.18) and Fig. 5.3, for sand $10^{-2} > \alpha > 10^{-3}\ \text{s}^{-1}$ and $1\ \text{m} < t_{50} < 10\ \text{m}$, hence the former condition applies and the amplitude response at all tidal frequencies is much reduced.

For silt and clay, $10^{-4} > \alpha > 10^{-6}\ \text{s}^{-1}$ and $2\ \text{h} < t_{50} < 200\ \text{h}$. Since the major tidal constituents lie in the range $\omega \geq 10^{-4}\ \text{s}^{-1}$, the cyclical amplitude response is relatively independent of $\alpha$ but proportional to tidal period, resulting in an enhancement of longer period constituents. Wherever there is a plentiful supply of erodible sediment of a wide size distribution, the resulting suspended sediment time series away from the immediate near-bed area is likely to be dominated by silt–clay.

In the absence of significant residual currents, the erosional time series for an $M_2$–$S_2$-dominated tidal regime will show pronounced components at $M_4$, $MS_4$, $MS_f$ and $Z_0$ frequencies, (5.4) and Table 5.1. These latter components are generated by non-linear combinations of $M_2$ and $S_2$ currents and not by any (usually small) tidal current amplitudes at these emergent frequencies. The similarity in amplitudes of the $M_4$ and $MS_4$ constituents may reduce the quarter-diurnal signal at neap tides (when their phases are opposed) and thereby suggest an enhanced semi-diurnal sediment signal that might be wrongly interpreted as indicating horizontal advection or a large diurnal current component. When 'residual currents' increase to a level of order 10% of the $M_2$ amplitude (as in strongly tidal shallow waters), the suspended sediment time series will include $M_2$ and $S_2$ constituents of comparable magnitude to those described for $M_4$, $MS_4$ and $MS_f$. Over a spring–neap tidal cycle, for $E > 10$ $W_S D$, the peak in suspended sediment concentration will generally occur 2–3 days after the occurrence of maximum tidal currents.

In summary, the theory (Table 5.1 and Fig. 5.6) explains how tidal spectra shown in SPM observations (Table 5.2 and Fig. 5.7) are modulated relative to that of the generating current spectra. The associated amplitude ratios and phase lags shown in such observations can be used to infer both the nature of the predominant sediment type and the roles of localised resuspension versus advection.

## Appendix 5A

### *5A.1  Analytical expression for sediment suspension*

An expression is derived for the time sequence of sediment suspension, linking erosion through to deposition as a function of three variables, namely settling velocity $W_S$, vertical dispersion coefficient $E$ and water depth $D$.

### *5A.2  Dispersion equation*

It is assumed that the distribution of suspended sediments can be described by the dispersion equation; moreover, in the present application, consideration is limited to motion in one horizontal direction, $X$. Thence, the version of (5.1) representing the rate of change in concentration $C$ is

$$\frac{dC}{dt} = \frac{\partial C}{\partial t} + U\frac{\partial C}{\partial X} + W\frac{\partial C}{\partial Z} = \frac{\partial}{\partial Z}E\frac{\partial C}{\partial Z} - \text{sinks} + \text{sources}, \qquad (5A.1)$$

where $t$ is time, $U$ horizontal velocity, $W$ vertical velocity, $Z$ the vertical axis measured upwards from the bed and $E$ a vertical dispersion coefficient. Horizontal dispersion is not considered, and $E$ is assumed to remain constant both temporally and vertically.

Equation (5A.1) can be converted to convenient variables as follows:

$$z = \frac{Z}{D}, \ w = \frac{W}{D}, \ e = \frac{E}{D^2},$$

where $D$ is the water depth, $z = 0$ is the bed and $z = 1$ the surface. By assuming horizontal homogeneity, i.e. $\partial C/\partial X = 0$, (5A.1) may be rewritten:

$$\frac{\partial C}{\partial t} + w\frac{\partial C}{\partial z} = e\frac{\partial^2 C}{\partial z^2} - \text{sinks} + \text{sources}. \tag{5A.2}$$

Equation (5A.2) can then be used to describe the transport of particulate material by setting the vertical velocity $W$ in (5A.1) equal to the settling velocity $-W_s$.

### 5A.3 Analytical solutions

Fischer *et al.* (1979) show that the general solution for dispersion of a tracer of mass $M$ (confined to 1D $Z$) released at $Z = 0$, $t = 0$ in a fluid moving at a velocity $W$ is

$$C(Z, t) = \frac{M}{\sqrt{4\pi Et}} \exp - \frac{(Z - Wt)^2}{4Et}. \tag{5A.3}$$

It is convenient to introduce variables $t' = w_s\, t$ and $z' = z + w_s\, t$. Rewriting (5A.3) in these variables, the concentration at $z'$ is

$$C(z', t) = \frac{M}{D\sqrt{4\pi \frac{e}{w_s} t'}} \exp\left(\frac{-z'^2}{4\frac{e}{w_s} t'}\right). \tag{5A.4}$$

Since (Fischer *et al.*, 1979),

$$\int\limits_0^z e^{-z^2}\,dz = \frac{\sqrt{\pi}}{2}\, \mathrm{ERF}(z), \tag{5A.5}$$

the net amount of sediment in suspension between $z' = 0$ and $z'$ is

$$TS = D\int\limits_0^{z'} C(z)\,dz = M'\, \mathrm{ERF}\left(\frac{z'}{\sqrt{4\frac{e}{w_s} t'}}\right), \tag{5A.6}$$

where $M' = M/2$ corresponds to the amount of eroded suspended above the bed at $t = 0$.

The value of the error function in (5A.6) must always be less than 1. Then, since for $0 < x < 1$, $\mathrm{ERF}(x) \sim x$, the net amount deposited at $z = 0$ after time $t'$ is given by substituting $z' = t'$ in (5A.6), i.e.

$$TS \approx \frac{M'}{2} \frac{t'}{\left(\frac{e}{w_s} t'\right)^{1/2}}. \tag{5A.7}$$

From (5A.7), after $t' = 0.04\ e/w_s$, $TS = 0.1M'$; after $t' = e/w_s$, $TS = 0.5M'$ while after $t' = 3.2\ e/w_s$, $TS = 0.9M'$. These deposits can be separated into an advective component due to $W_s$ in (5A.3) and a dispersive component associated with $E$. By successively differentiating the corresponding top and bottom terms in (5A.7) w.r.t. $t$, it follows that the advective deposition is a factor of 2 greater and of opposite sign to the dispersive 'erosion'. From differentiation of (5A.6) w.r.t. $z$, the rate of advective deposition is $W_S\ C_{(z'=0)}$, i.e. the anticipated net deposition for an absorptive bed that retains all particles making contact. The 50% return rate, implicit in (5A.3), infers a continuing dynamic relationship at the bed that persists up to some $4e/w_s$ after erosion. Wiberg and Smith (1985) used a 'rebound coefficient' of around 0.5 to account for partially elastic collisions between moving grains and the bed.

### 5A.4 Boundary condition at the surface z = 1

At the surface, a simple reflective boundary condition is introduced involving a ghost source (of equal magnitude to the original source) located at $z = 2 + w_s t$, i.e. equidistant from the original source at $z = -w_s\ t$. The 'first' (1% of $M'$) particles passing the surface corresponds to

$$\frac{1}{M'} \int_0^{D(1+w_s t)} C(Z)\ \mathrm{d}Z = 0.99 = \mathrm{ERF}\left(\frac{1 + t'}{\sqrt{4\frac{e}{w_s} t'}}\right). \tag{5A.8}$$

Equation (5A.8) requires the argument of the error function to equal 1.83, hence real solutions for $t'$ require $e/w_s > 0.3$. Thus, for smaller values of $e/w_s$, less than 1% of particles reach the surface.

### 5A.5 Boundary condition at the bed z = 2, 4, 6, etc.

Sediments 'reflected' from the surface arrive back at the bed where they may be deposited or resuspended so long as, by analogy with (5A.8), $(2 + t') / (4et' / w_s)^{1/2} < 1.83$, i.e. $e/w_s > 0.6$.

The ratio, $R$, of suspended sediments reflected from the surface towards the bed (i.e. between $z' = 1 + w_s t$ and $z' = 2 + w_s t$) to those yet to reach the surface (i.e. between $z' = w_s t$ and $z' = 1 + w_s t$) is

$$R = \frac{R_s}{R_B} = \frac{\mathrm{ERF}\left(\dfrac{2 + w_s t}{\sqrt{4et}}\right) - \mathrm{ERF}\left(\dfrac{1 + w_s t}{\sqrt{4et}}\right)}{\mathrm{ERF}\left(\dfrac{1 + w_s t}{\sqrt{4et}}\right) - \mathrm{ERF}\left(\dfrac{w_s t}{\sqrt{4et}}\right)}. \tag{5A.9}$$

This ratio shows that for $e/w_s < 0.1$ almost no particles reach the surface. Conversely, for $e/w_s > 10$, there will be little vertical variation in the suspended sediment concentrations.

Three approximations are considered for the bottom boundary condition as described below.

**[A] Fully reflective bed**

Sources of equal magnitude at

| INITIAL SOURCE $z = -w_s t$ | REFLECTION FROM BED | REFLECTION FROM SURFACE | |
|---|---|---|---|
| | | | **(1)** initial source |
| | | $(2 + w_s t)$ | **(2)** reflection of **(1)** |
| | $-(2 + w_s t)$ | | **(3)** reflection of **(2)** |
| | $\vert$ | | |
| | $\vert$ | | |
| | $-(2(n - 1) + w_s t)$ | | **(2n − 1)** reflection of **(2n − 2)** |
| | | $(2n + w_s t)$ | **(2n)** reflection of **(2n − 1)** |

Numerically, this series continues until the contribution from the last two terms is negligible. The condition implies that deposition only occurs at $z' = 0$, hence the solutions to (5A.6) shown by the dashed line in Fig. 5.2 indicate the rate of deposition.

**[B] Fully absorptive bed**

This requires zero effective concentration at the bed for those sediments arriving after reflection from the surface, i.e. sources of equal magnitude but opposite sign. This results in a series of sources/sinks as in boundary condition **[A]**, combinations 3, 4; 7, 8, etc. representing sinks.

**[C] Termination of the series after two terms**

Sources at

| INITIAL SOURCE $z = -w_s t$ | REFLECTION FROM BED | REFLECTION FROM SURFACE | |
|---|---|---|---|
| | | | **(1)** initial source |
| | | $(2 + w_s t)$ | **(2)** reflection of **(1)** |

Figure 5.2 illustrates the fraction of sediments remaining in suspension at time, $w_s t$, after erosion for these three boundary conditions. The three lines that correspond to the bottom boundary conditions **[A]** is dashed, **[B]** is dotted and **[C]** is solid.

For values of $e/w_s < 1.0$, the results show little difference since it was shown earlier that little sediment passes beyond $z = 2$ for this condition. However, for

$e/w_s > 10$, the results differ widely and, for the latter two boundary conditions, suspension times decrease with increasing values of $e/w_s$. This implies a greater 'capture rate' as a result of increased collisions with the bed.

As a convenient expedient, we subsequently adopt the third boundary condition as being someway intermediate between the other two extremes. However, it is recognised that this expedient reduces the validity of results presented for the range $e/w_s \gg 1$. In practice, the appropriate boundary condition will be a function of the benthic boundary layer dynamics and of the sediment characteristics and concentration.

Thus, subsequently, we assume sediment concentrations given by

$$C(z, t) = \frac{M}{D(4\pi et)^{1/2}} \left[ \exp - \frac{(z + w_s t)^2}{4et} + \exp - \frac{(2 + w_s t - z)^2}{4et} \right]. \quad (5A.10)$$

## References

Davies, A.G. and Thorne, P.D., 2008. Advances in the study of moving sediments and evolving seabeds. *Surveys in Geophysics*, doi:10.1007/S10712-008-9039-X, 36pp.

Fischer, H.B., List, E.J., Koh, R.C.Y., Imberger, J., and Brocks, N.K., 1979. *Mixing in Inland and Coastal Waters*. Academic Press, New York.

Gerritsen, H., Vos, R.J., van der Kaaij, T., Lane, A., and Boon, J.G., 2000. Suspended sediment modelling in a shelf sea (North Sea). *Coastal Engineering*, **41**, 317–352.

Grant, W.D. and Madsen, O.S., 1979. Combined wave and current interaction with a rough bottom. *Journal of Geophysical Research*, **84** (C4), 1797–1808.

Hutchinson, S.M. and Prandle, D., 1994. Siltation in the salt marsh of the Dee estuary derived from 137Cs analyses of short cores. *Estuarine, Coastal and Shelf Science*, **38** (5), 471–478.

Jones, S.E., Jago, C.F., Prandle, D., and Flatt, D., 1994. Suspended sediment dynamics, their measurement and modelling in the Dover Strait. In: (Beven, K.J., Chatwin, P.C., and Millbank, J.H. (eds), *Mixing and Transport in the Environment*, John Wiley and Sons, New York, pp. 183–202.

Krone, R.B., 1962. *Flume Studies on the Transport of Sediments in Estuarine Shoaling Processes*. Hydraulic Engineering Laboratories, University of Berkeley, CA.

Lane, A. and Prandle, D., 1996. Inter-annual variability in the temperature of the North Sea. *Continental Shelf Research*, **16** (11), 1489–1507.

Lane, A. and Prandle, D., 2006. Random-walk particle modelling for estimating bathymetric evolution of an estuary. *Estuarine, Coastal and Shelf Science*, **68** (1–2), 175–187.

Lavelle, J.W., Mojfeld, H.O., and Baker, E.T., 1984. An *in-situ* erosion rate for a fine-grained marine sediment. *Journal of Geophysical Research*, **89** (4), 6543–6552.

Partheniades, E., 1965. Erosion and deposition of cohesive soils. *Journal of Hydraulics Division ASCE*, **91**, 469–481.

Postma, H., 1967. Sediment transport and sedimentation in the estuarine environment. In: Lauff, G.H. (ed.), *Estuaries*. American Association for the Advancement of Science, Washington, DC, pp. 158–180.

Prandle, D., 1982. The vertical structure of tidal currents and other oscillatory flows. *Continental Shelf Research*, **1** (2), 191–207.

Prandle, D., 1997a. The dynamics of suspended sediments in tidal waters. *Journal of Coastal Research,* **40** (Special Issue No 25), 75–86.

Prandle, D., 1997b. Tidal characteristics of suspended sediment concentration. *Journal of Hydraulic Engineering*, ASCE, **123** (4), 341–350.

Prandle, D., 2004. Sediment trapping, turbidity maxima and bathymetric stability in tidal estuaries. *Journal of Geophysical Research*, **109** (C8).

Prandle, D., Murray, A. and Johnson, R., 1990. Analysis of flux measurements in the Mersey River. In: Cheng, R.T. (ed.), *Residual currents and Long Term Transport. Coastal and Estuarine Studies, Vol. 38*. Springer-Verlag, New York, pp. 413–430.

Romano, C., Widdows, J., Brimley, M.D., and Staff, F.J., 2003. Impact of Enteromorpha on near-bed currents and sediment dynamics: flume studies. *Marine Ecology Progress Series*, **256**, 63–74.

Sanford, L.P. and Halka, J.P., 1993. Assessing the paradigm of mutually exclusive erosion and deposition of mud, with examples from upper Chesapeake Bay. *Marine Geotechnics*, **114** (1–2), 37–57.

Soulsby, R.L., 1997. *Dynamics of Marine Sands: a Manual for Practical Applications*. Telford, London.

Van Rijn, L.C., 1993. *Principles of Sediment Transport in Rivers, Estuaries and Coastal Seas*. Aqua Publications, Amsterdam.

Van Rijn, L.C., 2007a. Unified view of sediment transport by currents and waves. 1: Initiation of motion, bed roughness and bed-load transport. *Journal of Hydraulic Engineering*, **133** (6), 649–667, doi:10.1061/(ASCE)0733-9429(2007)133:6(649).

Van Rijn, L.C., 2007b. Unified view of sediment transport by currents and waves. 2: Suspended transport. *Journal of Hydraulic Engineering*, **133** (6), 668–689, doi:10.1061/(ASCE)0733-9429(2007)133:6(668).

Van Rijn, L.C., 2007c. Unified view of sediment transport by currents and waves. 3: Graded beds. *Journal of Hydraulic Engineering*, **133** (7), 761–775, doi:10.1061/(ASCE)0733-9429(2007)133:7(761).

Wiberg, P.L. and Smith, J.D., 1985. A theoretical model for saltating grains in water. *Journal of Geophysical Research*, **90** (4), 7341–7354.

Winterwerp, J.C. and van Kesteren, W.G.M., 2004. *Introduction to the Physics of Cohesive Sediments in the Marine Environment. Developments in Sedimentology,* Vol. 56. Elsevier, Amsterdam, p. 466.

# 6

# Synchronous estuaries: dynamics, saline intrusion and bathymetry

## 6.1 Introduction

Previous chapters illustrated the nature of tidal elevations, currents, saline intrusion and sediment regimes in estuaries of varying sizes and shapes over a range of tidal, alluvial and river flow conditions. Here, we address the more fundamental question of how morphology is itself determined and maintained by the combined actions of tidal dynamics and the mixing of river and salt waters.

As in these earlier chapters, considerable simplifications are necessary to obtain generic analytical solutions to the governing equations. Here, we restrict consideration to strongly tidal, funnel-shaped 'synchronous' estuaries. Adopting the linearised 1D momentum and continuity equations, the focus is on propagation of a single predominant ($M_2$) tidal constituent. Since estuaries often involve significant differences in surface areas between low and high water, a triangular cross section is assumed.

Section 6.2 indicates how these approximations enable localised values for the amplitude and phase of tidal current $U^*$ to be determined in terms of depth, $D$ and elevation amplitude, $\varsigma^*$. A further expression for the slope of the sea bed, $SL$, enables both the shape and the length, $L$, of an estuary to be similarly determined.

Various derivations for the length of saline intrusion, $L_I$, were discussed in Chapter 4, based on the expedient assumption of a constant (in time and depth) axial density gradient $S_x$. All indicated a dependency on $D^2/fU^* U_0$, where $U_0$ is the residual river flow velocity and $f$ is the bed friction coefficient. Moreover, axial migration of the intrusion zone was recognised as vital in explaining observed variations in $L_I$ over the cycles of spring to neap tides and flood to drought river flows. The deduction, introduced in Section 6.3, that mixing occurs at a minimum in landward intrusion of salt is used here alongside expressions for $L$ and $L_I$ to derive an expression for $D_i$, the depth at the centre of the intrusion, in terms of $U_0$. This analysis indicates that $U_0$ is always close to $1\,\mathrm{cm\,s^{-1}}$, as commonly observed. Linking $U_0$ to river flow, $Q$, provides a morphological expression for depth at the mouth of an estuary as a function of $Q$ (with an additional dependence on side slope gradients).

In Section 6.4, the above results are converted into bathymetric Frameworks, mapping gross estuarine characteristics against $\varsigma^*$ and $Q$ (or $D$). In this way, a 'bathymetric zone' is postulated bounded by three conditions: $L_I/L < 1$, $E_x/L < 1$ ($E_x$ is the tidal excursion) alongside the Simpson-Hunter (1974) criteria for mixed estuaries $D/U^{*3} < 50\,\mathrm{m^2\,s^{-3}}$.

In Section 6.5, the validity of these new theories is assessed by comparison against observed bathymetries for 80 estuaries in England and Wales.

Peculiarly, the results derived in this chapter take no account of the associated sediment regimes. Chapter 7 considers the consequent implications for the prevailing sediment regime and rates of morphological adjustment.

## 6.2 Tidal dynamics

### 6.2.1 Synchronous estuary approach

In Chapter 2, it was shown how the essential characteristics of tidal dynamics in estuaries can be readily explained from analytical solutions. These dynamics are almost entirely determined by a combination of tides at the mouth and estuarine bathymetry with some modulation by bed roughness and, close to the head, river flows.

Dyer (1997) describes how frictional and energy conservation effects can combine in funnel-shaped bathymetry to produce a 'synchronous estuary' with constant tidal elevation amplitudes $\varsigma^*$. Prandle (2003) assessed the validity of the 'synchronous' solutions using numerical simulations. It was shown that for many convergent bathymetries, the assumption that $\partial\varsigma^*/\partial X \rightarrow 0$ remains valid except for a small section at the tidal limit. Moreover, it was shown in Fig. 2.5 that the shape and lengths derived for a synchronous estuary are in the centre of the observed range for funnel-shaped estuaries.

The present approach explores localised dynamics in terms of tidal elevation amplitude and water depth. The 'synchronous' assumption enables elevation gradients to be determined directly from tidal elevation amplitude. The solutions obtained are for a triangular-shaped estuary. Equivalent results for a rectangular cross section involve substitution of celerity $c = (gD)^{1/2}$ for the present result $c = (0.5\,gD)^{1/2}$. The dynamical solutions reduce to explicit functions of $\varsigma^*$, $D$ and bed friction coefficient $f$. The solutions are independent of the actual value of the side slopes; while their inclination can be asymmetric, they must remain (locally) constant. The analytical solutions assume that (i) tidal forcing predominates and can be approximated by the cross-sectionally averaged axial propagation of a single semi-diurnal tidal constituent; (ii) convective and density gradient terms can be neglected and (iii) the friction term can be sensibly linearised.

### 6.2.2 Analytical solution for 1D momentum and continuity equations (Prandle, 2003)

Omitting the convective term from the momentum equation, we can describe tidal propagation in an estuary by:

$$\frac{\partial U}{\partial t} + g\frac{\partial \varsigma}{\partial X} + f\frac{U|U|}{H} = 0 \tag{6.1}$$

$$B\frac{\partial \varsigma}{\partial t} + \frac{\partial}{\partial X} A U = 0, \tag{6.2}$$

where $U$ is velocity in the $X$-direction, $\varsigma$ is the water level, $D$ is the water depth, $H$ is the total water depth ($H = D + \varsigma$), $f$ is the bed friction coefficient (~0.0025), $B$ is the channel breadth, $A$ is the cross-sectional area, $g$ is the gravitational acceleration and $t$ is the time.

Concentrating on the propagation of one predominant tidal constituent ($M_2$), the solutions for $U$ and $\varsigma$ at any location can be expressed as

$$\varsigma = \varsigma^* \cos(K_1 X - \omega t) \tag{6.3}$$

$$U = U^* \cos(K_2 X - \omega t + \theta), \tag{6.4}$$

where $K_1$ and $K_2$ are the wave numbers, $\omega$ is the tidal frequency and $\theta$ is the phase lag of $U$ relative to $\varsigma$. The synchronous estuary assumption is that axial variations in $\varsigma^*$ are small. In deriving solutions to (6.1) and (6.2), a similar approximation is assumed to apply to $U^*$. The resulting solutions for $U^*$ (Fig. 6.1) indicate that this additional assumption is valid, except in the shallowest waters. Further assuming a triangular cross section with constant side slopes, (6.2) reduces to

$$\frac{\partial \varsigma}{\partial t} + U\left(\frac{\partial \varsigma}{\partial X} + \frac{\partial D}{\partial X}\right) + \frac{1}{2}\frac{\partial U}{\partial X}(\varsigma + D) = 0. \tag{6.5}$$

Friedrichs and Aubrey (1994) indicate that $U^*(\partial A/\partial X) \gg A(\partial U^*/\partial X)$ in convergent channels. Likewise assuming $(\partial D/\partial X) \gg (\partial \varsigma^*/\partial X)$, we adopt the following form of the continuity equation:

$$\frac{\partial \varsigma}{\partial t} + U\frac{\partial D}{\partial X} + \frac{D}{2}\frac{\partial U}{\partial X} = 0. \tag{6.6}$$

The component of $f U |U| /H$ at the predominant tidal frequency $M_2$ may be approximated by

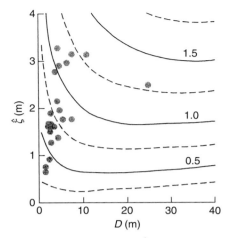

Fig. 6.1. Tidal Current amplitude, $U^*(\text{m s}^{-1})$, as $f$ $(D, \varsigma^*)$. From (6.9) with $f = 0.0025$.

$$\frac{8}{3\pi} \frac{25}{16} f \frac{|U^*|U}{D} = FU \qquad (6.7)$$

with $F = 1.33$ $fU^*/D$, where $8/3\pi$ derives from the linearisation of the quadratic friction term (Section 2.5). The factor $25/16$ is derived by assuming that the tidal velocity at any transverse location is given by a balance between the quadratic friction term (with localised depth) and a (transversally constant) surface slope – yielding a cross-sectionally averaged velocity of $4/5$ of the velocity at the deepest section.

Substituting solutions (6.3) and (6.4) into (6.1) and (6.6) with the frictional representation (6.7), four equations (pertaining at any specific location along an estuary) are obtained, representing components of $\cos(\omega t)$ and $\sin(\omega t)$ in (6.1) and (6.6). By specifying the synchronous estuary condition that the spatial gradient in tidal elevation amplitude is zero, we derive $K_1 = K_2 = k$, i.e. identical wave numbers for axial propagation of $\varsigma$ and $U$, thence

$$\tan\theta = -\frac{F}{\omega} = \frac{SL}{0.5\,D\,k}, \qquad (6.8)$$

where $SL = \partial D/\partial X$

$$U^* = \varsigma^* \frac{g\,k}{\left(\omega^2 + F^2\right)^{1/2}} \qquad (6.9)$$

$$k = \frac{\omega}{(0.5Dg)^{1/2}}. \qquad (6.10)$$

### 6.2.3 Explicit formulae for tidal currents, estuarine
### length and depth profile

A particular advantage of the above solutions is that they enable the values of a wide range of estuarine parameters to be calculated and illustrated as direct functions of $D$ and $\varsigma^*$. The ranges selected for illustration here are $\varsigma^*$ (0–4 m) and $D$ (0–40 m); these represent all but the deepest of estuaries.

*Current amplitudes* U*

Figure 6.1 (Prandle, 2004) shows the solution (6.9) with current amplitudes extending to $1.5\,\mathrm{m\,s}^{-1}$. As subsequently shown from Fig. 6.3, for $\varsigma^* \ll D/10$, these currents are insensitive to $f$. For $\varsigma^* \gg D/10$, these currents change by a factor of 2 over the range $f = 0.001–0.004$. Noting that for $F \gg \omega$, (6.9) indicates that $U^* \propto \varsigma^{*\,1/2}\, D^{1/4}\, f^{-1/2}$, thus illustrating why observed variations in $U^*$ are generally smaller than for $\varsigma^*$. Conversely, for $F \ll \omega$, (6.9) indicates that $U^* \propto \varsigma^*\, D^{-1/2}$. The contours show that maximum values of $U^*$ occur at approximately $D = 5 + 10\,\varsigma^*$; however, these are not pronounced maxima.

*Depth profile and estuarine length* L

In Fig. 6.2 (Prandle, 2004), utilising the values of $SL$ from (6.8), the length, $L$, of an estuary is calculated numerically by successively updating $SL$ as $D$ reduces along the estuary (assuming a constant value of $\varsigma^*$). By assuming $F \gg \omega$, an equivalent simple analytical solution can be determined

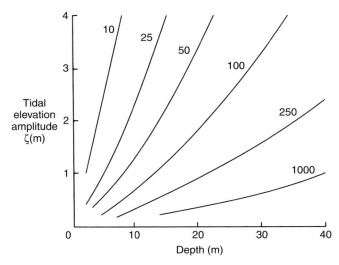

Fig. 6.2. Estuarine length, $L$ (km), as a $f(D, \varsigma^*)$, (6.12), with $f = 0.0025$.

$$D = \left( \frac{5}{4} \frac{(1.33 \, \varsigma^* f \omega)^{1/2}}{(2 \, g)^{1/4}} \right)^{4/5} x'^{\,4/5}, \qquad (6.11)$$

where $x' = L - X$.

Substituting $X = 0$ and $D = D_0$ at the mouth, estuarine lengths are given by

$$L = \frac{D_0^{5/4}}{\varsigma^*{}_0^{1/2}} \frac{4}{5} \frac{(2 \, g)^{1/4}}{(1.33 f \omega)^{1/2}} \sim 2460 \frac{D_0^{5/4}}{\varsigma^*{}_0^{1/2}} \qquad \text{for } f = 0.0025 \qquad (6.12)$$

(units m, subscripts 0 denote values at the mouth).

The dependency on $D^{5/4}/\varsigma^{*1/2}$ in (6.12) and Fig. 6.2 indicates that estuarine lengths are significantly more sensitive to $D$ than to $\varsigma^*$. Prandle (2003) shows that this expression for estuarine length is in broad agreement with data from some 50 estuaries (randomly selected using previously published data) located around the coasts of the UK and the eastern USA. For the UK estuaries, estimates of mud content were available, enabling some of the discrepancies between observed and estimated values of $L$ to be reconciled by introducing an expression for $f$ based on relative mud content.

### 6.2.4 Sensitivity to bed friction coefficient (f)

As indicated by (6.8) and (6.9), the sensitivities of the three parameters shown in Table 6.1 to the value of the bed friction coefficient $f$ depend on the value of $F/\omega$. The ratios of the change in parameters correspond to changes in bed friction coefficient $f' = \varepsilon f$. The effective extreme range of $\varepsilon$ is from 0.2 to 5. Prandle *et al.* (2001) showed how increased bed friction due to wave–current interaction reduced tidal current amplitudes by up to 70%.

Figure 6.3 (Prandle, 2004) illustrates that $F/\omega$, the ratio of the friction to inertial terms determined from (6.8), is approximately equal to unity for $\varsigma^* = D/10$. For values of $\varsigma^* \gg D/10$, tidal dynamics become frictionally dominated, whereas for $\varsigma^* \ll D/10$ friction becomes insignificant. Friedrichs and Aubrey (1994) showed the predominance

Table 6.1 *Sensitivity to bed friction coefficient* f' = ε f

|  |  | $F/\omega \gg 1$ or $\varsigma^* \gg D/10$ | $F/\omega \ll 1$ or $\varsigma^* \gg D/10$ |
|---|---|---|---|
| Current amplitude | $U$ | $\varepsilon^{-1/2}$ | 1 |
| Seabed slope | $SL$ | $\varepsilon^{1/2}$ | $\varepsilon$ |
| Estuarine length | $L$ | $\varepsilon^{-1/2}$ | $\varepsilon^{-1}$ |

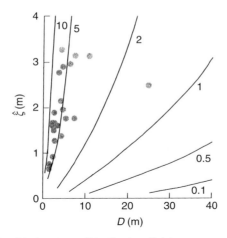

Fig. 6.3. Ratio of the friction term $F$ to the inertial term $\omega$ as a $f(D, \varsigma^*)$, (6.8).

of the friction term in convergent channels, irrespective of depth – this accords with large values of $SL$ in (6.8).

### 6.2.5 Rate of funnelling in a synchronous estuary

Chapter 2 described generalised frameworks for tidal response in funnel-shaped estuaries based on analytical solutions for (6.1) and (6.2). Using the bathymetric approximations of depth proportional to $X^m$ and breadth to $X^n$, Prandle and Rahman (1980) produced a response framework (Fig. 2.5), showing relative tidal amplitude response and associated phases across a wide range of estuarine bathymetries. Here, we calculate how the estuarine lengths and shapes derived for synchronous estuaries fit within this framework. The 'funnelling factor' $v$ is given by

$$v = \frac{n+1}{2-m}. \tag{6.13}$$

Since (6.11) corresponds to $m=n=0.8$, the synchronous estuary solution corresponds to $v = 1.5$, i.e. close to the centre of values encountered.

The vertical axis in Fig. 2.5 represents a transformation of the distance from the head of the estuary given by

$$y = \frac{4\pi}{2-m} \left( \frac{X}{(gD)^{1/2}P} \right)^{(2-m)/2}, \tag{6.14}$$

where $P$ is the tidal period. Hence for $X=L$, $m=0.8$, from (6.12)

$$y = 0.9 \left( \frac{D^{3/4}}{\varsigma^{*1/2}} \right)^{0.6}. \tag{6.15}$$

Taking $D = 5$, $\varsigma^* = 4$ (m) together with $D = 20$, $\varsigma^* = 2$ (m) as representative of minimum and maximum values of $y$ in 'mixed' estuaries, these correspond to $y = 1.22$ and $y = 2.8$. Figure 2.5 shows that this range of values for $y$ is also representative of observed lengths, extending from a small fraction to almost a quarter wavelength for the $M_2$ constituent.

## 6.3 Saline intrusion

We restrict interest to mixed or partially mixed estuaries and assume a, temporally and vertically, constant relative axial density gradient, $S_x = (1/\rho)(\partial\rho/\partial X)$, with density linearly proportional to salinity. In Chapter 4, the following expression, (4.44), for saline intrusion length, $L_I$, in mixed estuaries was derived (Prandle, 1985):

$$L_I = \frac{0.005 \, D^2}{f \, U^* \, U_0}. \tag{6.16}$$

Here we link the above to a further expression determining the location of the intrusion along the estuary to derive salient characteristics within the intrusion zone. In conjunction with the results from Section 6.2, this provides an expression for the depth at the mouth of an estuary as a function of river flow.

### 6.3.1 Stratification levels and flushing times

The earlier derivation of tidal current amplitudes (6.9) enables direct estimation of the Simpson and Hunter (1974) criterion $D/U^{*3} > 50 \, \text{m}^2 \, \text{s}^{-3}$ for stratification to persist. The results, shown in Fig. 6.4 (Prandle, 2004), indicate that this implies that estuaries with tidal elevation amplitudes $\varsigma^* > 1$ m will generally be mixed.

An additional indication of stratification levels can be calculated from the time, $T_K$, for complete vertical mixing by diffusion of a point source – estimated by Prandle (1997) as $T_K = D^2/K_Z$ ($K_Z$ is the vertical diffusivity). Approximating $K_Z = f U^* D$ yields $T_K = D/f U^*$, estimates of which are shown in Fig. 6.5 (Prandle, 2004). This approach produces results consistent with those in Fig. 6.4. Noting that, for semi-diurnal constituents, whenever $T_K > P/2 \sim 6$ h, stratification is likely to persist beyond consecutive peaks of mixing on flood and ebb tides. Conversely for $T_K < 1$ h, little stratification is likely. For intermediate values, $1$ h $< T_K < 6$ h, intra-tidal stratification is likely – especially via tidal straining on the flood tide (Simpson *et al.*, 1990).

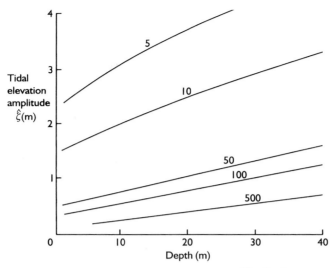

Fig. 6.4. Simpson–Hunter stratification parameter $D/U^{*3}$ m$^2$ s$^{-3}$.

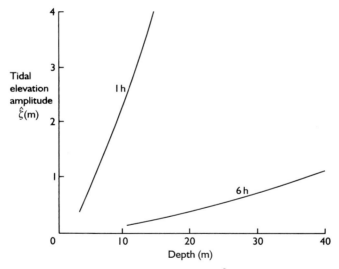

Fig. 6.5. Time for vertical mixing by diffusion $D^2/K_z$, $K_z = fU^*D$ with $U^*$ from (6.9).

An estimate of salinity flushing rates based on the time for river water to replace half of the salinity content of an estuary is given by

$$T_F = \frac{0.5\,(L_I/2)}{U_0} = \frac{0.0013\,D^2}{f\,U^*\,U_0^2}. \tag{6.17}$$

### 6.3.2 *Location of mixing zone, residual current associated with river flow*

In Chapter 4, estimates of the landward limits of saline intrusion $x_u = (X_i - L_1/2)/L$ corresponding to successive values of $x_i = (X_i/L)$, the centre of the intrusion, were compared with observations in eight estuaries. The best agreement occurred when the landward limit of saline intrusion was a minimum.

Adopting this latter result as a criterion to determine the position, $x_i$, where the saline intrusion will be centred, requires in dimensionless terms

$$\frac{\partial}{\partial x}(x - 0.5\, l_i) = 0. \qquad (6.18)$$

Substituting $l_i = L_1/L$, utilising (6.16) and (6.12) and introducing the shallow water approximation to (6.9), yields

$$U^{*2} = \frac{\varsigma^* \, \omega \, (2\,g\,D)^{1/2}}{1.33\, f}. \qquad (6.19)$$

Then assuming $Q = U_0 D_i^2/\tan \alpha$, where $\tan \alpha$ is the side slope of the triangular cross section, we obtain

$$x_i^2 = 333 \,\frac{Q \tan \alpha}{D_0^{5/2}}. \qquad (6.20)$$

Noting that the depth, $D_i$, at $x_i$ is $D_0\, x_i^{0.8}$, we obtain

$$U_0 = \frac{D_i^{1/2}}{333} \,\mathrm{m\,s}^{-1}. \qquad (6.21)$$

For depths ranging from a few metres to tens of metres, (6.21) yields values of $U_0$ close to $1\,\mathrm{cm\,s}^{-1}$, as commonly observed. Noting that (6.20) corresponds to $l_i = 2/3x_i$, these values for $U_0$ will increase by a factor of 2 at the upstream limit and decrease by 40% at the downstream limit.

If we introduce estuarine bathymetry of the form breadth $B_0 x^n$ and depth $D_0 x^m$, we obtain the following alternative form for (6.20):

$$x_i = \left(\frac{855\,Q}{D_0^{3/2} B_0\,(11m/4 + n - 1)}\right)^{1/(11m/4+n-1)}. \qquad (6.22)$$

An especially interesting feature of the results for the axial location of saline intrusion, (6.20) and (6.22), and the expression for residual river flow current, (6.21), is their independence of both tidal amplitude and bed friction coefficient, although there is an implicit requirement that tidal amplitude is sufficient to maintain partially mixed conditions. Equations (6.20) and (6.22) emphasise how the centre of the intrusion adjusts for changes in river flow $Q$. This 'axial migration' can severely complicate the sensitivity of saline intrusion beyond the anticipated direct responses apparent from the expression (6.16) for the length of intrusion, $L_I$.

## 6.4  Estuarine bathymetry: theory

### 6.4.1  Morphological zone determined by tidal dynamics and stratification

Using the above result that the riverine component of velocity in the saline intrusion region approximates $1 \ \mathrm{cm \ s}^{-1}$, Fig. 6.6 (Prandle, 2004) illustrates typical values of the lengths of saline intrusion obtained from (6.16). Moreover, combining this latter result with that for estuarine length, $L$, (6.12), Fig. 6.7 (Prandle, 2004) shows the ratio $L_I/L$.

Similarly, Fig. 6.8 (Prandle, 2004) shows the ratio of tidal excursion $E_x$, (6.24), as a fraction of $L$ for a tracer released at the mouth on the flood tide. These values for $E_x$ include compensation for the reduction in tidal velocity with decreasing (upstream depths) but ignore axial variations in phase.

By introducing requirements for saline mixing to be contained within an estuary, we can define a 'bathymetric zone' bounded by

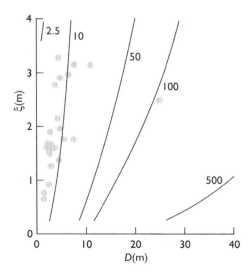

Fig. 6.6. Saline intrusion length, $L_I$ (km), (6.16). Values scale by $0.01/U_0$, $U_0$ in $\mathrm{m \ s}^{-1}$.

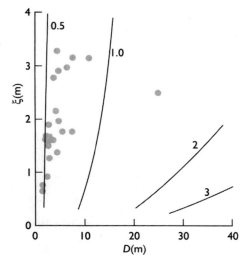

Fig. 6.7. Ratios of saline intrusion to estuarine length, $L_{\mathrm{I}}/L$, (6.16) and (6.12).

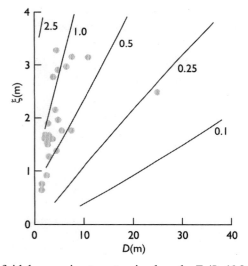

Fig. 6.8. Ratios of tidal excursion to estuarine length, $E_{\mathrm{x}}/L$, (6.24) and (6.12).

$$\text{(i)} \quad \frac{L_{\mathrm{I}}}{L} < 1,$$

$$\text{(ii)} \quad \frac{E_{\mathrm{x}}}{L} < 1 \text{ and}$$

$$\text{(iii)} \quad \frac{D}{U^{*3}} < 50 \,\mathrm{m}^{-2}\,\mathrm{s}^3, \tag{6.23}$$

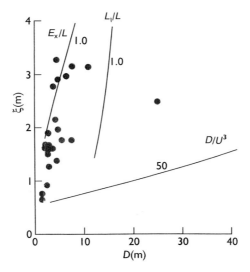

Fig. 6.9. Zone of estuarine bathymetries. Bounded by $E_x < L$, $L_I < L$ and $D/U^{*3} < 50\,\mathrm{m^2\,s^{-3}}$, (6.23).

where the tidal excursion

$$E_x = \left(\frac{2}{\pi}\right) U^* P. \tag{6.24}$$

This zone, shown in Fig. 6.9 (Prandle, 2004), is reasonably consistent with the superimposed distribution of $(D, \varsigma^*)$ values from 25 UK estuaries (Prandle, 2003).

### 6.4.2 Estuarine depths as a function of river flow

The results for $x_i$ and $D_i$, (6.21) and (6.22) in Section 6.3.2, can be used to obtain estimates of the depth $D_0$ at the mouth of the estuary. Noting that with $l_i = 2/3\ x_i$, for the intrusion to be confined to the estuary, the maximum value for $x_i = 0.75$. Inserting this value for $x_i$ into (6.21), we obtain

$$D_0 = 12.8\ (Q \tan \alpha)^{0.4}. \tag{6.25}$$

Combining this result with (6.12), the estuarine length, $L$, is given by

$$L = 2980 \left(\frac{Q \tan \alpha}{f \varsigma^*}\right)^{1/2}. \tag{6.26}$$

Where estuarine bathymetries were established under historical conditions with much larger (glacial melt) values of $Q$, we might expect saline mixing to start landwards of the mouth. Conversely, where saline mixing involves an offshore plume, we postulate either exceptionally large values of $Q$ or that bathymetric erosion to balance existing river flow is hindered.

The results for $U_0$, (6.21), and $D_0$, (6.25), are independent of both the friction coefficient, $f$, and the tidal amplitude, $\varsigma^*$. O'Brien (1969) noted that the minimum flow area of tidal inlets was effectively independent of the type of bed material. However, the two expressions for estuarine length, (6.12) and (6.26), are dependent on the inverse square root of both $f$ and $\varsigma^*$.

### *Observed versus computed estuarine bathymetries*

Examination of a range of UK estuaries indicated that in general, $0.02 > \tan \alpha > 0.002$, (6.25) then corresponds to

$$2.68\, Q^{0.4} > D_0 > 1.07\, Q^{0.4}. \tag{6.27}$$

Figure 6.10 shows results from UK, USA and European estuaries (Prandle, 2004). For the steeper side slope, (6.27) yields values for $D_0$ of 2.7 m for $Q = 1\,\mathrm{m^3\,s^{-1}}$; 6.7 m for $Q = 10\,\mathrm{m^3\,s^{-1}}$; 16.9 m for $Q = 100\,\mathrm{m^3\,s^{-1}}$ and 42.4 m for $Q = 1000\,\mathrm{m^3\,s^{-1}}$. Comparable figures for the smaller side slope are depths of 1.1, 2.7, 6.7 and 16.9 m. Figure 6.10 shows that the envelope described by (6.27) encompasses almost all of the observed estuarine co-ordinates of $(Q, D)$.

The mean discharge of the world's largest river, the Amazon, is $200\,000\,\mathrm{m^3\,s^{-1}}$, representing 20% of net global freshwater flow. Moreover, the cumulative discharge

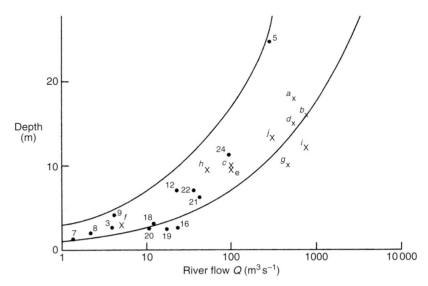

Fig. 6.10. Depth at the mouth as a function of river flow. Theoretical envelope (6.27), observed values from Prandle (2004). UK estuaries are labelled by numbers, others by letters.

of the next nine largest rivers amounts to a similar total (Schubel and Hirschberg, 1982). Outside of these ten largest rivers, $Q < 15\,000\,\mathrm{m}^3\,\mathrm{s}^{-1}$, from (6.25) this corresponds to $D = 50\text{--}125\,\mathrm{m}$. Thus, the range of values shown in Fig. 6.10 clearly represents the vast majority of estuaries. Moreover, we note from Fig. 6.7 that the larger estuaries with $D > 10\,\mathrm{m}$ will often involve freshwater plumes extending seawards.

## 6.5 Estuarine bathymetry: assessment of theory against observations

The above theories for estuarine bathymetry provide formulations for

(1) depth at the mouth, $D$ versus river flow, $Q$, (6.25);
(2) tidal length $L$ versus $D$ and $\varsigma^*$ (tidal amplitude), (6.12);
(3) a bathymetric zone, delineated on a framework of $D$ and $\varsigma^*$ (Fig. 6.9).

A morphological data set (Future-Coast, Burgess *et al.*, 2002) for 80 estuaries in England and Wales, Fig. 6.11(Buck and Davidson, 1997), was used to assess the validity of the above theories. The UK estuaries include large inter-tidal zones, with breadths at high tide typically three or more times low tide values. Hence, as for the earlier theoretical developments, the following analyses of observed morphologies assume triangular cross sections with side-slope $\tan\alpha = 2D/B$. To separate results for differing geomorphological types in this assessment, we use the same morphological classifications as Buck and Davidson (1997), i.e. Rias, Coastal Plain and Bar-Built. For these estuarine types, mean observed values of the estuarine parameters – $D$, $\varsigma^*$, $Q$, breadth $B$ and side-slope $\tan\alpha$ – are shown in Table 6.2.

### 6.5.1 Statistical analyses

The observational data set extends over a diverse range of estuaries with attendant uncertainties and inaccuracies introduced both during the original data collection and in the subsequent syntheses of the data. All statistical fits removed two outliers as determined by iterative calculations. The subsequent analyses are sub-divided into four categories: (i) all estuaries, (ii) Rias, (iii) Coastal Plain and (iv) Bar-Built. The numbers, $N$, in each category are as follows: 80(28), 15(61), 30(45) and 35(42) with the bracketed figures indicating percentage significant correlation at the 99% level.

An estimate of percentage of variance (PVA) accounted for is used where

$$\mathrm{PVA} = 100 \times \left(1 - \left(\sum |O_i^2 - X_i^2|\right) \Big/ \sum O_i^2\right) \qquad (6.28)$$

with $O_i$ observed and $X_i$ the 'best' statistical fit over the range of estuaries $i = 1-N$. The use of normalised PVA is appropriate for the large parameter ranges involved. The statistical fit is assumed to take the form $y = Ax^P$, with $A$ calculated from least squares over the range $-3 < P < 3$ (in increments of 0.01) and with $P$ chosen to reflect the maximum value for PVA.

Fig. 6.11. Estuaries of England and Wales.

Table 6.2 *Observed estuarine lengths, L, depths, D, river flows, Q, and side-slope, tan α*

| Type | No. | $L \sim AD^P$ | R (PVA) | Mean L (km) | $D \sim AQ^P$ | R (PVA) | Mean D (m) | Mean Q (m³ s⁻¹) | Mean (tan α) |
|------|-----|----------------|---------|-------------|----------------|---------|-----------|------------------|---------------|
| All | 80 | $1.28D^{1.24}$ | 0.69 | 20 | $3.3Q^{0.47}$ | 0.55 | 6.5 | 14.9 | 0.013 |
| Ria | 15 | $0.99D^{1.10}$ | 0.89 | 12 | $5.1Q^{0.32}$ | 0.74 | 9.3 | 6.3 | 0.037 |
| Coastal Plain | 30 | $1.95D^{1.12}$ | 0.69 | 33 | $3.0Q^{0.38}$ | 0.67 | 8.1 | 17.9 | 0.011 |
| Bar-built | 35 | $1.92D^{1.15}$ | 0.66 | 9 | $2.4Q^{0.35}$ | 0.72 | 3.6 | 9.5 | 0.014 |
| Theory | | $\mathbf{1.83D^{1.25}}$ | | | $\mathbf{2.3Q^{0.40}}$ | | | | |

*Note:* Best statistical Fits and correlation (PVA, percentage of variance accounted for). Theoretical values are from (6.12) and (6.25) with mean values $\varsigma^* = 1.8$ m and tan $\alpha = 0.013$. *Source:* Prandle *et al.*, 2006.

The numerical optimisation calculates $r^{1/q}$ versus $s^q$, where $r$ and $s$ are any fitted parameter and $q = P^{1/2}$, this ensures that the fit applies when the order of the parameters is reversed.

Mean values for river flow velocity, $U_0$, were calculated from the values for $Q \tan \alpha / D^2$ shown in Table 6.2. These are as follows (in cm s⁻¹): all estuaries, 0.5; Rias, 0.04; Coastal Plain, 0.3 and Bar-Built, 1.0, i.e. consistent with estimates from (6.21). Prandle (2004) showed that values of $U_0$ derived both from observations worldwide and from numerical model calculations are generally in the range 0.2–1.5 cm s⁻¹. The much lower values for Rias reflect their peculiar morphological development.

### 6.5.2 Theory versus observed morphologies

The summary of observed estuarine bathymetries shown in Table 6.2 usefully encapsulates the descriptions of estuarine types outlined by Davidson and Buck (1997). Thus, in general, Rias are short, deep and steep-sided with small river flows. Coastal Plain estuaries are long and funnel-shaped with gently sloping triangular cross sections providing extensive inter-tidal zones. Bar-Built estuaries are short and shallow with small values of both river flow and tidal range. Prandle (2003) noted that sandy estuaries tend to be short and muddy estuaries tend to be long. In sedimentary terms, Bar-Built estuaries are located along coasts with plentiful supplies of marine sediments and, consequently, are close to present-day equilibrium. Coastal Plain estuaries are continuing to infill following 'over-deepening' via post-glacial river flows while Rias are drowned river valleys (with related cross sections) as a consequence of (relative) sea level rise.

*Length (*$L \sim AD^P$*) and depth* (D)

The powers, 1.12 and 1.15 of $D$, shown in Table 6.2 for Coastal Plain and Bar-Built estuaries lie close to the theoretical value of 1.25 as do the coefficients 1.95 and 1.92 to the theoretical 1.83. The power, 1.10, of $D$ for Rias is in reasonable agreement with the theoretical value of 1.25, but the reduced value 0.99 of the coefficient reflects the shorter lengths of these estuaries.

Overall, we note statistically significant relationships between all of these estuarine parameters in all types of estuaries indicating the tendency for estuarine morphologies to be confined within restricted parameter ranges.

*Depth (*$D \sim AQ^P$*) and river flow* (Q)

The powers of $Q$, 0.32 for Rias and 0.38 for both Coastal Plain and Bar-Built estuaries, are all close to the theoretical value of 0.40. Likewise, the related values for the coefficient $A$ are close to the theory except for Rias where the higher coefficient reflects their greater depths.

The relationships between $L$ and $D$ shown in Table 6.2 can be used to estimate the power $m$ for the Prandle and Rahman (1980) estuarine response, Fig. 2.5, in which $D \propto x^m$. Prandle (2006) extended these statistical analyses to calculate corresponding relationships between $L$ and breadth, $B$, thereby providing estimates of $n$ for $B \propto x^n$. Together, from (6.13), these values of $m$ and $n$ indicate the following values of the 'funelling factor' $v$: All, 1.85; Ria, 1.72; Coastal Plain, 2.07 and Bar-Built, 2.25. Maximum tidal amplification occurs for $v = 1$ with considerable reduction of this peak for $v > 2$. Thus, tidal elevations and currents are likely to be more spatially homogeneous in Bar-Built estuaries reflecting conditions closer to equilibrium.

*Bathymetric zone*

Figure 6.12 (Prandle *et al.*, 2005) shows the 'zone of estuarine bathymetries' bounded by (6.23). This encompasses most of the observed estuaries. For $\varsigma^* > 3$ m, Fig. 6.1 shows that tidal current amplitudes can exceed $1.5 \text{ m s}^{-1}$ and the axial slope of the bed increases significantly. Consequently, such estuaries are generally in the form of deep contiguous bay-estuary systems such as the Bristol Channel or Bay of Fundy.

### 6.5.3 *Minimum depths and flows for estuarine functioning*

Figures 6.4 and 6.5 indicate that $\varsigma^* > 1$ m for a 'mixed' estuary, this translates to a minimum value of $D = 1$ m and from (6.12) $L \sim 2.5$ km, for an estuary to function over a complete tidal cycle.

This minimum length requirement corresponds to $D > \varsigma^{*0.4}$. Then, by substituting the values for transverse slope, $\tan \alpha$, from Table 6.1 into (6.25) we derive minimum values of $Q$ in $\text{m}^3 \text{ s}^{-1}$ for $\varsigma^* = 1$, 2 and 4 m as shown in Table 6.3.

Table 6.3 *Minimum values of river flow,* Q *(m³ s⁻¹), for estuaries*
*to function over a complete tidal cycle of amplitude* ς

|  | $\varsigma^* = 1$ m | $\varsigma^* = 2$ m | $\varsigma^* = 4$ m |
|---|---|---|---|
| All estuaries | 0.13 | 0.75 | 4.2 |
| Rias | 0.05 | 0.26 | 1.5 |
| Coastal Plain | 0.15 | 0.88 | 5.0 |
| Bar-Built | 0.12 | 0.69 | 3.9 |

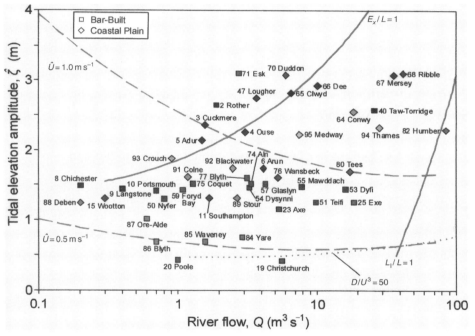

Fig. 6.12. Bathymetric zone. Bounded by: $E_x < L$, $L_1 < L$ and $D/U^{*3} < 50$ m² s⁻³.

### 6.5.4 Spacing between estuaries

Having postulated that estuarine bathymetry is determined by $\varsigma^*$ and $Q$, the question
arises as to how estuaries adjust over geological time scales to climate change. In
particular, what are the consequences of changes in rainfall and catchment areas as
coasts advance or retreat under falling or rising msl. These questions are addressed
further in Chapter 8; here we distil these questions into estimates of the spacing
between estuaries.

For a long straight coastline with spacing, $SP$ (km), between estuaries and a rectangular catchment of landward extent $CT$ (km), the river flow $Q$ is given by

$$Q = 0.032\, SP \times CT \times R, \qquad (6.29)$$

where $R$ is the annual rainfall reaching the river (m a$^{-1}$). Thus, typical UK values of $SP=10$ km, $CT=50$ km and $Q=15$ m$^3$ s$^{-1}$ indicate that $R\sim0.9$ m a$^{-1}$, which is in broad agreement with observations.

By introducing (6.25) with $\tan\alpha = 0.013$ (Table 6.2), we obtain the following expression for spacing between estuaries:

$$SP = 41\, \frac{D^{5/2}}{R \times CT}. \qquad (6.30)$$

We note from Table 6.2 and from global observational data shown by Prandle (2004) that few estuaries have values of $D > 20$ m. Hence, to avoid small values of $SP$ for continental land masses with large values of $CT$, we anticipate the formation of deltas or multiple 'sub-estuaries' linked to the sea by tidal basins such as in Chesapeake Bay.

## 6.6 Summary of results and guidelines for application

By introducing the assumption of a 'synchronous' estuary (where the surface slope due to the gradient in phase of tidal elevations significantly exceeds that from changes in amplitude $\varsigma^*$), explicit localised expressions are obtained for both the amplitude and phase of tidal currents and the slope of the sea bed, $SL$, in terms of $(D, \varsigma^*)$ where $D$ is water depth. Integration of the expression for $SL$ provides an estimate of the shape and length, $L$, of an estuary. By further combining these results with (4.44) for the length of saline intrusion, $L_I$, an expression linking the depth at the estuarine mouth with river flow, $Q$, is derived, i.e. a morphological framework linking the size and shape of any estuary to river flow and tidal amplitude.

Recent attempts to forecast how estuarine bathymetries may evolve as a consequence of Global Climate Change have prompted the more fundamental question:

### *What determines estuarine shape, length and depth?*

Estuaries form an interface between the rise and fall of the tide at the coast and the river discharges. Hence, we expect bathymetries must reflect a combination of tidal amplitude, $\varsigma^*$, and river flow, $Q$, alongside some representation of the alluvium. However, geological adjustment rates are extremely slow in comparison with the (relatively) rapid changes in msl (and hence water depth) and $Q$. Thus, we anticipate that bathymetries reflect some intermediate adjustment between antecedent formative conditions and subsequent present-day dynamic equilibrium. This adjustment rate will depend on both the supply of sediments for deposition and the 'hardness' of the geology for erosion.

### Tidal currents and frictional influence

By introducing the 'synchronous' estuary assumption, Section 6.2 derives localised expressions, (6.8) and (6.9), for the amplitude and phase of tidal currents in terms of $\varsigma^*$ and $D$, illustrating why $U^*$ is invariably in the range $0.5 < U^* < 1.5 \, \mathrm{m \, s}^{-1}$ (Fig. 6.1). These synchronous solutions also explicitly quantify the issue raised in Chapter 2 concerning the ratio of the frictional to inertial terms. Figure 6.3 shows how this ratio approximates $10 \, \zeta^* : D$.

### Lengths and shape

The expression (6.8) for the slope of the sea bed, $SL$, provides an estimate of the shape of an estuary (6.11). By subsequent integration of these slopes, Fig. 6.2 shows estuarine lengths, $L$ (6.12), as a function of $\varsigma^*$ and $D$ and the bed friction coefficient, $f$, where the latter introduces some representation of the alluvium.

### Range of estuarine morphologies

By combining these results for $U^*$ and $L$ with the expression (6.16) for the length of saline intrusion, $L_I$, a 'bathymetric zone' for 'mixed' estuaries is determined enveloped by the conditions, $L_I/L < 1$, $E_x/L < 1$ ($E_x$ tidal excursion length) and the Simpson-Hunter (1974) criterion for vertical mixing ($D/U^{*3} < 50 \, \mathrm{m}^{-2} \, \mathrm{s}^{3}$). Figures 6.5 and 6.6 use the expression (6.9) for $U^*$ to indicate how the latter criterion for determining mixed versus stratified estuaries closely approximate the condition $\varsigma^* \sim 1 \, \mathrm{m}$.

By introducing the observation that mixing occurs at a minimum in landward intrusion of salt, an expression linking the depth at the mouth of an estuary $D_0$ with river flow, $Q$, is derived, (6.21) and (6.25). Interestingly, this expression is independent of both $\varsigma^*$ and $f$. Assuming a triangular cross section with side slopes, $\tan \alpha$, between 0.02 and 0.002, Fig. 6.10 illustrates the resultant envelope of $D_0$ as a function of $Q$. This envelope supports the conclusion, derived in Chapter 4, that the river flow velocity within the saline intrusion zone in mixed estuaries is invariably of the order of $(1 \, \mathrm{cm \, s}^{-1})$.

In Section 6.5, it is shown that for 'mixed' estuaries to function over the complete tidal cycle requires minimum values of $D \approx \varsigma^{*0.4}$ and $Q \approx 0.25 \, \mathrm{m}^3 \, \mathrm{s}^{-1}$. Typical maximum values of around $D \approx 20 \, \mathrm{m}$ and $\varsigma^* \approx 3 \, \mathrm{m}$ can be postulated and used to suggest the formation of deltas and composite estuaries in the draining of large continental land masses.

## *Assessment of bathymetric framework*

Using the above relationship between $D_0$ and $Q$, the earlier 'bathymetric zone' is converted into a framework for estuarine morphology in terms of the 'boundary conditions' $\varsigma^*$ and $Q$, quantifying how tides and river flows determine estuarine size and shape. This framework, Fig. 6.12, is assessed against an observational data set extending to 80 UK estuaries.

While individual estuaries exhibit localised features (related to underlying geology, flora and fauna, historical development and 'intervention'), the overall values of depth, length and (funnelling) shape are shown to be consistent with the new dynamical theories. Moreover, the morphological features which characterise Rias, Coastal Plain and Bar-Built estuaries can be rationalised by reference to these theories.

The derived theoretical proportionality of $L$ to $D_0^{1.25}$ was substantiated for all three estuarine types. For both Coastal Plain and Bar-Built estuaries, magnitudes of $L$ are in close agreement with the theory while Rias indicate significantly reduced values for $L$.

Similarly, the theoretical proportionality of $D_0$ to $Q^{0.4}$ was substantiated for all three estuarine types, with good magnitude agreement for Coastal Plain and Bar-Built estuaries while Rias indicated significantly larger values of $D$.

## *Theoretical Frameworks*

Results derived in this chapter are synthesised into comprehensive Theoretical Frameworks describing estuarine dynamics, salinity intrusion and bathymetry in terms of the 'natural' boundary conditions $Q$, $\varsigma^*$ and $f$. The underlying relationships are as follows.

| | | | |
|---|---|---|---|
| (a) Current amplitude | $U^* \propto \varsigma^{*\frac{1}{2}} D^{\frac{1}{4}} f^{-\frac{1}{2}}$ | Shallow water | (6.9) |
| | $\propto \varsigma^* D^{-1/2}$ | Deep water | |
| (b) Estuarine length | $L \propto D^{5/4}/\varsigma^{*\frac{1}{2}} f^{\frac{1}{2}}$ | | (6.12) |
| (c) Depth variation | $D(x) \propto x^{0.8}$ | | (6.11) |
| (d) Depth at the mouth | $D_0 \propto (aQ)^{0.4}$ | | (6.25) |
| (e) Ratio of friction: inertia | $F/\omega \propto 10\varsigma^*/D$ | | (6.8) |
| (f) Stratification limit | $D/U^{*3} \sim \varsigma^* = 1\,\mathrm{m}$ | | (6.24) |
| (g) Salinity intrusion | $L_\mathrm{I} \propto D^2/fU_0U^*$ | | (6.16) |
| (h) Bathymetric zone | Bounded by $L_\mathrm{I} < L$, $E_x < L$ | | (6.23) |
| | and $D/U^{*3} < 50\,\mathrm{m}^{-2}\mathrm{s}^3$ | | |
| (i) Flushing time | $T_\mathrm{F} \propto L_\mathrm{I}/U_0$ | | (6.17) |

These Frameworks provide perspectives against which to assess the morphology of any particular estuary. By identifying 'anomalous' estuaries, possible causes can be explored such as distinct regional patterns of historical evolution, engineering 'interventions', 'hard' geology, dynamics or mixing inconsistent with the theoretical assumptions, non-representative observational data, vagaries of sediment supply and wave impact.

The impact of flora and fauna has long been recognised as influencing the supply, consolidation and bio-turbation of bed sediments. Section 6.2 shows how associated variations in the effective bed friction factor, $f$, have a major impact on dynamics and thence on both morphology and the sediment regime.

## *Corollary*

The validation of these new theories provokes a paradigm shift, explored in Chapter 7, suggesting that prevailing estuarine sediment regimes are the consequence of rather than the determinant for bathymetries.

## References

Buck, A.L. and Davidson, N.C., 1997. *An Inventory of UK Estuaries. Vol. 1. Introduction and Methodology.* Joint Nature Conservancy Committee, Peterborough, UK.

Burgess, K.A., Balson, P., Dyer, K.R., Orford, J., and Townend, I.H., 2002. Futurecoast the integration of knowledge to assess future coastal evolution at a national scale. In: *28th International Conference on Coastal Engineering. American Society of Civil Engineering*, Vol. 3, Cardiff, UK, pp. 3221–3233.

Davidson, N.C. and Buck, A.L., 1997. *An inventory of UK estuaries. Vol. 1. Introduction and Methodology.* Joint Nature Conservation Committee, Peterborough, UK.

Dyer, K.R., 1997. *Estuaries: a Physical Introduction*, 2nd ed. John Wiley, Hoboken, NJ.

Friedrichs, C.T. and Aubrey, D.G., 1994. Tidal propagation in strongly convergent channels. *Journal of Geophysical Research*, **99** (C2), 3321–3336.

O'Brien, M.P., 1969. Equilibrium flow area of inlets and sandy coasts. *Journal of Waterways and Coastal Engineering Division ASCE*, **95**, 43–52.

Prandle, D., 1985. On salinity regimes and the vertical structure of residual flows in narrow tidal estuaries. *Estuarine Coastal and Shelf Science*, **20**, 615–633.

Prandle, D., 1997. The dynamics of suspended sediments in tidal waters. *Journal of Coastal Research*, **25**, 75–86.

Prandle, D., 2003. Relationships between tidal dynamics and bathymetry in strongly convergent estuaries. *Journal of Physical Oceanography*, **33** (12), 2738–2750.

Prandle, D., 2004. How tides and river flows determine estuarine bathymetries. *Progress in Oceanography*, **61**, 1–26.

Prandle, D., 2006. Dynamical controls on estuarine bathymetry: assessment against UK data base. *Estuarine Coastal and Shelf Science*, **68** (1–2), 282–288.

Prandle, D. and Rahman, M., 1980. Tidal response in estuaries. *Journal of Physical Oceanography*, **70** (10), 1552–1573.

Prandle, D., Lane, A., and Manning, A.J., 2006. New typologies for estuarine morphology. *Geomorphology*, **81** (3–4), 309–315.

Prandle, D., Lane, A., and Wolf, J., 2001. Holderness coastal erosion – Offshore movement by tides and waves. In: Huntley, D.A., Leeks, G.J.J., and Walling, D.E. (eds), *Land–Ocean Interaction, Measuring and Modelling Fluxes from River Basins to Coastal Seas*. IWA publishing London, pp. 209–240.

Simpson, J.H. and Hunter, J.R., 1974. Fronts in the Irish Sea. *Nature*, **250**, 404–406.

Simpson, J.H., Brown, J., Matthews, J., and Allen, G., 1990. Tidal straining, density currents and stirring in the control of estuarine stratification. *Estuaries*, **13** (2), 125–132.

Schubel, J.R. and Hirschberg, D.J., 1982. The Chang Jiang (Yangtze) estuary: establishing its place in the community of estuaries. In: Kennedy, V.S. (ed.), *Estuarine Comparisons*. Academic Press, New York, pp. 649–654.

# 7

# Synchronous estuaries: sediment trapping and sorting – stable morphology

## 7.1 Introduction

Suspended sediments in estuaries generally increase upstream to produce a 'turbidity maximum' (TM) in the vicinity of the upstream limit of saline intrusion, with concentrations hugely increased relative to open-sea conditions. Uncles *et al.* (2002) summarise observational studies of these *TM* and discuss mechanisms responsible for their formation. Here, for the case of strongly tidal estuaries, we develop generic quantitative expressions to represent the mechanisms producing these high sediment concentrations. The aim is to identify the scaling parameters which determine the sensitivities to sediment type (sand to clay), spring to neap tides and drought to flood river flows. Recognising the long-term stability of estuarine bathymetry, despite the continuous large ebb and flood sediment fluxes, an additional aim is to identify feedback processes that maintain this stability.

Postma (1967) first described the mechanisms responsible for estuarine trapping of fine sediments, namely gravitational circulation, non-linearities in the tidal dynamics and delays between resuspension and settlement. However, these respective roles are difficult to isolate either from observations or from models. Hence, it has been difficult to gain a clear insight into the scaling of these processes and to estimate their sensitivities to changes in either marine or fluvial forcing or to internal parameters.

### 7.1.1 Earlier studies

Postma (1967) described the nature of sediment distributions in tidal estuaries and indicated likely controlling mechanisms. Salient characteristics identified by Postma include

(1) concentrations in suspension of fine sediments that are much higher than in related marine or fluvial sources;

(2)  provenance, both of suspended and of surficial bed sediments, is overwhelmingly of marine origin;

(3)  evidence of lags between peaks in current and concentration, up to 4 days for the spring–neap cycle;

(4)  lags that are associated with cumulative erosion and delayed settling for finer particles;

(5)  lags that are negligible for particles with diameters $d \geq 100\,\mu m$ (settling velocity $W_s \geq 0.01\,\text{m}\,\text{s}^{-1}$) and

(6)  peak suspended concentrations (*TM*) are widely observed, often related to the upstream limit of gravitational circulation (saline intrusion) and with particle sizes typically $100 > d > 8\,\mu m$.

Postma states that while estuaries may contain both coarse and fine material, it is the characteristics of the latter which generally predominate in determining bathymetry in conjunction with tidal amplitude, river flows and sediment supplies. Postma also notes the impacts of both flocculation and waves on sediment regimes. While these latter impacts are widely acknowledged, the details of these processes are not considered here.

Dronkers and Van de Kreeke (1986) observed that even in tidal bays with no effective river flow, a landward increase in SPM is often observed. They showed how Postma's (1967) concepts of ebb–flood asymmetry can be quantified to determine the net transport of fine sediments.

Festa and Hansen (1978), using a 2D laterally averaged model, indicated how *TM* could be generated by gravitational circulation. Their study effectively assumed a rectangular cross section and involved tidally averaged dynamics. Using a steady-state numerical model with river flow $U_R = 0.02\,\text{m}\,\text{s}^{-1}$ (omitting both erosion and deposition), they found that the strength and location of the *TM* depended critically on settling velocity, with pronounced *TM* for $W_s > 5 \times 10^{-6}\,\text{m}\,\text{s}^{-1}$. Dyer and Evans (1989) incorporated sequential erosion and deposition into a vertically averaged estuarine model and showed how the specification of the erosion threshold influenced the balance of import and export of suspended sediments.

The widely observed occurrence of *TM* close to the upstream limit of saline intrusion led to assumptions that gravitational circulation was the generating mechanism. Jay and Musiak (1994) emphasised the roles of both barotropic and baroclinic mechanisms and the need to incorporate axial and vertical components of these in estimating net sediment fluxes. Jay and Kukulka (2003) quantified net residual sediment fluxes associated with coupling between tidal variations in current and SPM. They show how these fluxes can far exceed the fluxes derived from a product of net residual current and tidally averaged concentration. Their analysis is similar to the present one, except that for partially stratified systems, baroclinic modifications to tidal current profiles are more significant.

Markofsky *et al.* (1986) showed that the *TM* in the Weser traps silt, acting as a filter within the estuary. Utilising observed time series of suspended sediments in the Weser, Grabemann and Krause (1989) showed that localised deposition and resuspension are dominant processes in the *TM*, with systematic repetition of conditions over disparate cycles of tides and river flows. Lang *et al.* (1989) developed a 3D numerical model to reproduce these observations, incorporating a settling velocity $W_s = 5 \times 10^{-4} \, \mathrm{m \, s^{-1}}$. They suggested that some correlation might exist between patterns of energy dissipation and source distributions of suspended sediments.

Hamblin (1989) indicated that local resuspension, ebb–flood asymmetry and saline intrusion all contribute to *TM* in the St. Lawrence River. A model simulation specified $W_s = 3 \times 10^{-4} \, \mathrm{m \, s^{-1}}$, consistent with observed particle diameters in the range $10 < d < 20 \, \mu\mathrm{m}$.

Uncles and Stephens (1989) describe extensive measurement of *TM* in the Tamar River, where tidal range varies between 2 and 6 m. They noted an order of magnitude difference in *TM* concentrations over the spring–neap cycle. They also noted a preponderance of silt $20 < d < 40 \, \mu\mathrm{m}$, both suspended and deposited within the *TM* region. They suggested that resuspension, tidal pumping and gravitational circulation may all be contributing mechanisms. The authors also note the importance of flocculation of fine sediments in suspension and of the rapid change in density/porosity of recently deposited sediments.

In a subsequent investigation of *TM* in the Tamar River, Friedrichs *et al.* (1998) determined that *TM* were primarily linked to internally generated non-linearities with three dominant effects: (i) flood-dominant asymmetry, (ii) river flow and (iii) settling lag effected via breadth convergence. Conversely, axial variations in depth were not found to be important. An axially varying bed erodability was found to be necessary to maintain an overall sediment budget. Their model introduces a time lag between peaks in current speed and concentration of 45 min.

Aubrey (1986) emphasised the role of tidal distortions associated with channel geometry and bed friction in producing flood or ebb dominance in shallow estuaries. Aubrey noted the dependence of the magnitude of these distortions on $\varsigma^*/D$. Here, we concentrate on estuaries with large tidal amplitude–depth ratios in shallow, triangular cross sections with associated strong vertical mixing.

### 7.1.2 Approach

The present approach involves direct integration of expressions for tidal dynamics, salinity intrusion and sediment motions into an analytical emulator.

The emulator is applicable within strongly tidal (hence mixed) funnel-shaped estuaries and incorporates processes that are pronounced in shallow estuaries with triangular cross sections. It provides clear illustrations of parameter dependencies and enables conditions of zero net sediment flux to be determined. While these results are, to some degree, dependent on a priori assumptions, the approach has the advantage of (i) relatively straightforward mathematics, (ii) generic application over a wide range of parameters, namely tidal elevation amplitude $\varsigma^*$, water depth $D$, river flow $Q$, sediment type or fall velocity $W_s$ and friction coefficient $f$.

The primary objective is to integrate into an 'analytical emulator' simplified descriptions of the salient mechanisms of

(1) tidal and residual currents associated with saline intrusion, river flow and tidal non-linearities;
(2) sediment erosion, suspension and deposition

to produce expressions for mean concentrations and net cross-sectional fluxes of SPM.

Section 7.2 summarises results for (1) as described previously in Chapters 4 and 6. Similarly, Section 7.3 summarises results for (2), based on Chapter 5, introducing continuous functional representations (7.22) for the half-life of sediments in suspension, $t_{50}$, and for their vertical profile $e^{-\beta z}$, (7.24). In Section 7.4, the validity of the resulting expressions for sediment concentrations and net fluxes are assessed against numerical model simulations (Fig. 7.1; Prandle, 2004c).

Section 7.5 examines component contributions to net sediment flux with attendant sensitivity analyses. It is shown how separate current components from the tidal non-linearities, involving $\cos\theta$ and $\sin\theta$, largely determine the balance between import and export, where $\theta$ is the phase difference between tidal elevation, $\varsigma^*$ and current, $U^*$. The combinations of parameter values which produce zero net sediment fluxes are then determined.

Section 7.6 summarises these earlier results indicating in Fig. 7.8 how, for fall velocities of 0.0001, 0.001 and 0.01 m s$^{-1}$, the balance between net import or export changes as tidal amplitudes vary from 1 to 4 m (indicative of neap to spring variations). Proceeding upstream from deep to shallow water, the balance between import of fine sediments and export of coarser ones becomes finer, i.e. selective 'sorting' and trapping. Likewise, more imports, extending to a coarser fraction, occur on spring than on neap tides.

In Section 7.7, the above results are encapsulated into new estuarine typology framework (Figs 7.9–7.11) comparing observed versus theoretical dynamics, bathymetry and sediment regimes for a wide range of estuaries.

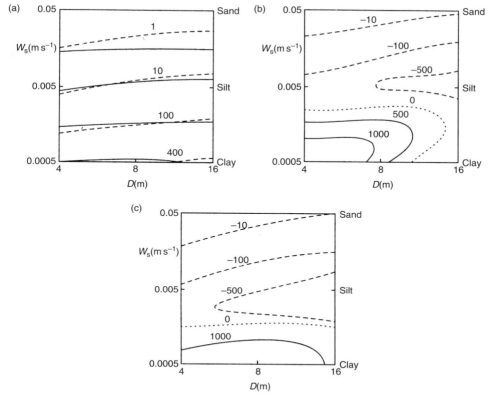

Fig. 7.1. Analytical Emulator versus Numerical Model values of suspended concentrations and net fluxes.
Model (solid contour), analytical emulator (7.29a) (dashed);
(a) Depth- and time-averaged concentrations (mg l$^{-1}$),
(b) net depth-integrated sediment fluxes from the numerical model;
(c) net depth-integrated sediment fluxes from the analytical emulator (7.31).
Values are for $\zeta^* = 2$ m and $f = 0.0025$. Fluxes are in t m$^{-1}$ width per year, positive landward.

## 7.2 Tidal dynamics, saline intrusion and river flow

Estuarine dynamics are largely determined by tides and estuarine bathymetry with some modulation by river flow evident close to the tidal limit. For the case of a synchronous estuary, Chapter 6 presents localised solutions for the amplitude and phase of tidal currents expressed in terms of tidal elevation amplitude and water depth. For completeness, these solutions are summarised here. Similarly, the salient features of the dynamics associated with saline intrusion and river flow, described in Chapter 4, are summarised.

These solutions apply to the case of a single (predominant) tidal constituent in a triangular-shaped estuary. Convective and density gradient terms are neglected, and the friction term is linearised. It can be shown (Prandle, 2004b) that tidal propagation in 'mixed' estuaries is almost entirely unaffected by saline intrusion.

### 7.2.1  Analytical solution for 1D tidal propagation
### (Chapter 6; Prandle, 2003)

Omitting the convective term from the momentum equation, we can describe tidal propagation in an estuary by

$$\frac{\partial U}{\partial t} + g\frac{\partial \varsigma}{\partial X} + f\frac{U|U|}{H} = 0 \tag{7.1}$$

$$B\frac{\partial \varsigma}{\partial t} + \frac{\partial}{\partial X} A\, U = 0, \tag{7.2}$$

where $U$ is velocity in the $X$ direction, $\varsigma$ is water level, $D$ is water depth, $H$ is the total water depth ($H = D + \varsigma$), $f$ is the bed friction coefficient ($\sim 0.0025$), $B$ is the channel breadth, $A$ is the cross-sectional area, $g$ is gravitational acceleration and $t$ is time.

Concentrating on the propagation of one predominant tidal constituent, M$_2$, the solutions for $U$ and $\varsigma$ at any location can be expressed as

$$\varsigma = \varsigma^* \cos(K_1 X - \omega t) \tag{7.3}$$

$$U = U^* \cos(K_2 X - \omega t + \theta), \tag{7.4}$$

where $K_1$ and $K_2$ are the wave numbers, $\omega$ is the tidal frequency and $\theta$ the phase lag of $U^*$ relative to $\varsigma^*$.

Further assuming a triangular cross section with constant side slopes, where $U\partial A/\partial X \gg \partial U/\partial X$ and $\partial D/\partial X \gg \partial \varsigma/\partial X$, we simplify the continuity equation to

$$\frac{\partial \varsigma}{\partial t} + U\frac{\partial D}{\partial X} + \frac{D}{2}\frac{\partial U}{\partial X} = 0. \tag{7.5}$$

The component of $fU|U|/H$ at the predominant tidal frequency M$_2$ may be approximated by (Prandle, 2004a):

$$\frac{8}{3\pi}\frac{25}{16} f\frac{|U^*|U}{D} = FU, \tag{7.6}$$

with $F = 1.33\, f\, U^*/D$. By specifying the synchronous estuary condition that the spatial gradient in tidal elevation amplitude is zero, we derive $K_1 = K_2 = k$, i.e. identical wave numbers for axial propagation of $\varsigma$ and $U$. Thus,

$$\tan\theta = \frac{F}{\omega} = -\frac{SL}{0.5\, D\, K} \tag{7.7}$$

$$U^* = \varsigma^* \frac{g\,k}{(\omega^2 + F^2)^{1/2}},$$  (7.8)

where $SL = \partial D / \partial X$, and

$$k = \frac{\omega}{(0.5Dg)^{1/2}}.$$  (7.9)

Thus, in shallow water, values of $U^*$ are proportional to $D^{1/4}(\varsigma^*/f)^{1/2}$ and $\tan\theta$ is proportional to $D^{-3/4}(\varsigma^* f)^{1/2}$.

### 7.2.2 Saline intrusion (Chapter 4)

For the case of a well-mixed estuary with a constant axial salinity gradient, $S_x$, the following expression for residual velocities associated with saline intrusion, $U_s$, at fractional height $z$ ($=Z/D$) above the bed was derived using (4.15), Prandle (1985):

$$U_s = g\,S_x \frac{D^3}{K_z}\left(\frac{-z^3}{6} + 0.269z^2 - 0.0373z - 0.0293\right),$$  (7.10)

where $K_z$ is the vertical eddy diffusivity.

For well-mixed conditions, the length of saline intrusion $L_I$ is given by (4.44):

$$L_I = \frac{0.005\,D}{f\,U^*\,U_R},$$  (7.11)

where $U_R$ is the residual velocity component associated with river flow and $S_x$ may be approximated by the 'relative' salinity gradient:

$$S_x = \frac{0.027}{L_I}.$$  (7.12)

Here we assume that the eddy diffusivity coefficient equals the eddy viscosity coefficient and is given by

$$K_Z = E = f\,U^*\,D.$$  (7.13)

Then from (7.10), (7.11), (7.12) and (7.13), we estimate $U_s$ at the bed as

$$U_S = -1.55\,U_R,$$  (7.14)

independent of $f$, $U^*$ or $D$. Moreover, it was shown in Chapter 5 that $U_R$ is generally order of $(1\text{ cm s}^{-1})$; hence, we anticipate values of residual currents associated with an axial salinity gradient of typically 1 or 2 cm s$^{-1}$. Prandle (2004b) showed that the neglect of convective overturning in deriving (7.10) results in an underestimation of $U_s$ by a factor of up to 2. Figures 6.4 and 6.5 (Prandle, 2004a) show that for a synchronous estuary, stratification is generally confined to $\varsigma^* < 1$ m. However, some element of intra-tidal stratification will occur for larger values of $\varsigma^*$. Such stratification can significantly increase values of $U_s$; for example, Dronkers and Van de Kreeke (1986) show values of $U_s$ up to $10 \, \text{cm} \, \text{s}^{-1}$ in the partially stratified Volkerak estuary.

### 7.2.3  River flow

Equation (4.12) provides the following approximation for the vertical profile of river flow in a strongly tidal estuary:

$$U_R = 0.89 \, \bar{U}_R \left( -\frac{z^2}{2} + z + \frac{\pi}{4} \right). \tag{7.15}$$

## 7.3  Sediment dynamics

Chapter 5 describes localised solutions for suspended sediment concentrations. These are summarised here prior to incorporation alongside synchronous estuary solutions in Section 7.4.

### 7.3.1  Suspension, erosion, deposition and vertical profiles

Neglecting horizontal components of advection and dispersion, a localised distribution of suspended sediments can be described by the dispersion equation

$$\frac{\partial C}{\partial t} - W_S \frac{\partial C}{\partial Z} = K_Z \frac{\partial^2 C}{\partial Z^2} - S_k + S_c, \tag{7.16}$$

where $S_k$ and $S_c$ are sinks and sources of sediment. Following Prandle (1997a and 1997b), we assume sediment concentration time series associated with each erosional event of magnitude $M$, (5.11):

$$C(Z,t) = \frac{M}{(4\pi K_z t)^{1/2}} \left[ \exp - \frac{(Z + W_s t)^2}{4 K_z t} + \exp - \frac{(2D + W_s t - Z)^2}{4 K_z t} \right]. \tag{7.17}$$

This assumes deposition at the bed both by advection and dispersion for particles that return to the bed following 'reflection' at the surface. An observed time series represents a time integration of all such preceding 'events'.

### Erosion

Erosion $ER(t)$, due to a tidal current $U(t)$, is assumed to be of the form

$$ER(t) = f\gamma\rho|U|^N, \tag{7.18}$$

where $N$ is some power of velocity typically in the range of 2–5 and $\rho$ is water density. For $N=2$, Prandle *et al.* (2001) derived $\gamma = 0.0001 \, \text{m}^{-1}$ s; for convenience, this formulation is adopted here. This implies a rate of erosion directly proportional to frictional bed stress.

### Deposition

Chapter 5 describes solutions of (7.16), showing that deposition can be approximated by an exponential function $e^{-\alpha t}$ with 'half-life' $t_{50} = 0.693/\alpha$, where $\alpha$ is given by the larger of (5.14) and (5.15), i.e.

$$\alpha = 0.693 \, \frac{W_{\text{s}}^2}{K_Z} \tag{7.19}$$

$$\alpha = 0.1 \, \frac{K_z}{D^2}. \tag{7.20}$$

The parameter $K_z/W_{\text{s}}D$, which characterises the governing mechanics, is inversely proportional to the familiar Rouse number. It is shown in Chapter 5 that maximum half-lives occur at $K_z/W_{\text{s}}D = 2.5$. For $K_z/W_{\text{s}}D \gg 1$, dispersion predominates, and sediments are well mixed vertically. Conversely, for $K_z/W_{\text{s}}D \ll 1$, advective settling predominates, and sediments remain close to the bed. Figure 5.4 shows the relationship between $W_{\text{s}}$ and $K_z/W_{\text{s}}D$ for a range of values of $\varsigma^*$ (utilising (7.8) and (7.13)). This illustrates how the demarcation line of $K_z/W_{\text{s}}D \sim 1$ coincides with fall velocities of order of $(1 \, \text{mm s}^{-1})$ or, for

$$W_{\text{s}}(\text{ms}^{-1}) = 10^{-6} \, d^2(\mu) \tag{7.21}$$

a particle diameter $d$ of order of $(30 \, \mu\text{m})$. It is shown in Section 7.5 that the sediment balance in estuaries is especially sensitive to sediments in this range.

To construct an analytical emulator, we need a continuous transition between conditions (7.19) and (7.20). By simple curve fitting, the results shown in Fig. 5.2 can be closely approximated by (5.17):

$$\alpha = \frac{0.693 \, W_{\text{s}}/D}{10^x}, \tag{7.22}$$

where $x$ is the root of the equation

$$x^2 - 0.79x + j(0.79 - j) - 0.144 = 0 \tag{7.23}$$

with $j = \log_{10} K_z/W_{\text{s}}D$.

### Vertical profile of suspended sediments

In subsequent estimates of net tidal fluxes, continuous functional descriptions of the vertical profiles of suspended sediment concentrations are required. By numerical fitting of a profile $e^{-\beta z}$ to simulations based on (7.17), the following expression for $\beta$ was derived using (5.19):

$$\beta = \left[ 0.91 \log_{10} \left( 6.3 \, \frac{K_z}{W_s D} \right) \right]^{-1.7} - 1. \tag{7.24}$$

Figure 5.5 shows, for a range of values of $K_z/W_sD$, values of $\beta$ from (7.24) alongside related sediment profiles. These results illustrate how effective complete vertical mixing is achieved for $K_z/W_sD > 2$, and 'bed load' only occurs for $K_z/W_s D < 0.1$.

## 7.4 Analytical emulator for sediment concentrations and fluxes

Here, the dynamical results described in Section 7.2 are integrated with the sediment solutions from Section 7.3 to form an 'analytical emulator' for determining net sediment fluxes (Prandle, 2004c). Erosion, assumed proportional to velocity squared, is modulated by an exponential settling rate to yield mean and tidally varying components of sediment concentration. The approach assumes that continuous cyclical erosion and deposition coexist without threshold values. Figure 7.7 summarises how these flux components are determined from the product of tidal and residual velocities (modified for vertical structure components) with these constituents of sediment concentration (likewise modified for vertical structure).

### 7.4.1 Erosional velocity components

Section 2.6.3 shows that to maintain continuity in a tidally varying cross section where net flux is $U_1$ times the mean cross-sectional area, the $M_2$ current $U_1^* \cos(\omega t)$ given by (7.4) must be accompanied by both $M_4$ and $Z_0$ (residual) components:

$$\begin{aligned} U_2^* &= -aU_1^* [\cos(2\omega t - \theta)] \\ U_0' &= -aU_1^* \cos \theta, \end{aligned} \tag{7.25}$$

where $a = \varsigma^*/D$ for a triangular channel and $a = 0.5\,\varsigma^*/D$ for a rectangular channel.

### Mean concentration

Henceforth for clarity of other symbols, the asterisks used to designate tidal amplitudes, $\varsigma^*$ and $U^*$, are omitted.

Assuming erosion is proportional to velocity $V$ squared, the expansion of $M_2$, $M_4$ and $Z_0$ velocity constituents gives

$$V^2 = [U_1 \cos \omega t + U_2 \cos (2\omega t - \theta) + U_0]^2, \tag{7.26}$$

where $U_0 = U_0' + U_R + U_s$, with $U_2$ and $U_0'$ from (7.25) and $U_R$ and $U_s$ from (7.15) and (7.10), respectively. This yields suspended mass components at the following tidal frequencies (scaled by $f\gamma\rho$ from (7.18)):

$$(U_1 \cos \omega t + U_2 \cos(2\omega t - \theta) + U_0)^2 = V_0^2 + V_\omega^2 + V_{2\omega}^2 + V_{3\omega}^2 + V_{4\omega}^2$$

with $V_0^2 = 0.5(U_1^2 + U_2^2) + U_0^2$

$$
\begin{aligned}
V_\omega^2 &= U_1 U_2 \cos(\omega t - \theta) + 2 U_0 U_1 \cos \omega t \\
V_{2\omega}^2 &= 2 U_0 U_2 \cos(2\omega t - \theta) + 0.5 U_1^2 \cos 2\omega t \\
V_{3\omega}^2 &= U_1 U_2 \cos(3\omega t - \theta) \\
V_{4\omega}^2 &= 0.5 U_2^2 \cos(4\omega t - 2\theta).
\end{aligned}
\tag{7.27}
$$

### 7.4.2 Mass of sediments in suspension

From Section 5.5, the mass in suspension for each cyclical erosion component at frequency $\omega$, when modulated by deposition at the rate $e^{-\alpha t}$, is given by (5.20):

$$C(t) = \int_{-\infty}^{t} \cos \omega t' e^{-\alpha(t-t')} \, dt' = \frac{\alpha \cos \omega t + \omega \sin \omega t}{\alpha^2 + \omega^2}, \tag{7.28}$$

where the integral sums backward in time to represent all remaining contributions to suspended concentration.

Thus, (7.28) modulates the erosion at any frequency to yield a concentration proportional to $1/(\alpha^2 + \omega^2)^{1/2}$. From (7.27), (7.28) and (7.18), we note that the net mass in suspension includes components at frequencies $\sigma$ as follows:

$$\sigma = 0, \text{ mass MCI} \quad CD = f\gamma\rho \frac{[U_0^2 + 0.5(U_1^2 + U_2^2)]}{\alpha} \tag{7.29a}$$

$$\sigma = \omega, \text{ mass MC2} \quad CD = f\gamma\rho \frac{U_1 U_2}{\alpha^2 + \omega^2}[\alpha \cos(\omega t - \theta) + \omega \sin(\omega t - \theta)] \tag{7.29b}$$

$$\sigma = \omega, \text{ mass MC3} \quad CD = f\gamma\rho \frac{2 U_0 U_2}{\alpha^2 + \omega^2}[\alpha \cos \omega t + \omega \sin \omega t]. \tag{7.29c}$$

Additional components can be shown not to contribute to net sediment flux and, hence, are omitted in (7.29).

Figure 7.1a (Prandle, 2004c) shows the corresponding mean concentrations obtained from (i) a cyclical tidal numerical model simulation of (7.17) and (ii) the $\sigma=0$ component of (7.29). Good agreement is shown over a range of values of both $W_s$ and $D$. These results are for $\varsigma^* = 2$ m, $f=0.0025$ and $U_R=0.01$ m s$^{-1}$, with $U_1$ calculated from (7.8).

The full concentration time series can be calculated by summing each of the components, $\sigma = 0$, $\omega$ to $4\omega$, in (7.27) with their modulation by deposition at the rate $e^{-\alpha t}$ as indicated in (7.28).

### 7.4.3 Net sediment fluxes

Sediment fluxes involve the product of erosion, from (7.27), modulated by deposition, via (7.28), multiplied by the velocity component appropriate to an (assumed) constant depth i.e. $U_1 \cos \omega t + U_{RS}$, where $U_{RS} = U_R + U_S$. The vertical structure of concentrations is approximated by $e^{-\beta z}$, with $\beta$ estimated from (7.24). The resulting concentration time series can be calculated by summing each of the five components in (7.27) together with their modulation by deposition at the rate $e^{-\alpha t}$ (Section 7.3).

The vertical structure of the velocity components associated with salinity intrusion and river flow are specified from (7.10) and (7.15), respectively. The vertical structure for the $M_2$ tidal constituent is approximated by

$$U(z) = \bar{U}\left(0.7 + 0.9\,z - 0.45\,z^2\right), \tag{7.30}$$

where $\bar{U}$ is the depth-averaged tidal velocity amplitude.

No account of vertical phase variations is considered. For the $M_4$ constituent, it can be shown that only the ratio of bed to depth mean value is required; for simplicity, a uniform vertical structure is assumed.

The parameters $U_1$, $U_2$ and $U_0$, describing erosion in (7.26), represent velocities at the bed. We introduce the coefficients $P$ and $Q$ to represent depth variations (relative to the bed) in $U_1$ and $U_{RS}$, respectively. From (7.30) the value of $P \sim 1/0.7$, while from (7.14) and (7.15) $Q \sim 1/-0.69$. Net sediment fluxes are then obtained by multiplying the erosional components (7.27), modulated via the suspension expression (7.28), and the vertical profile (7.24) by the velocity component ($PU_1 \cos \omega t + QU_{RS}$). The latter is consistent with the assumption of a net semi-diurnal water flux plus river and salinity components. This yields residual sediment fluxes ($F$) as follows:

$$F1 = \frac{1}{\alpha}\,Q\,U_{RS}\left[U_0^2 + 0.5\,(U_1^2 + U_2^2)\right] \tag{7.31a}$$

for mass MC1 and flow $QU_{RS}$.

$$F2 = \frac{1}{\alpha^2 + \omega^2}\,P\,U_1^2\,U_2\,(0.5\,\alpha \cos \theta - 0.5\,\omega \sin \theta) \tag{7.31b}$$

for mass MC2 and flow $PU_1 \cos \omega t$.

$$F3 = \frac{1}{a^2 + \omega^2} \, P \, U_1^2 \, U_0 \, \alpha \qquad (7.31c)$$

for mass MC3 and flow $PU_1 \cos \omega t$.

For (7.31), flux magnitudes are again scaled by $f\gamma\rho$, and concentrations are modulated by the vertical structure of the concentration $\beta e^{-\beta z}/(1 - e^{-\beta})$.

The first component, F1 (mass component MC1), represents the flux associated with the product of (i) time-averaged SPM concentration, MC1 in (7.29), and (ii) residual currents associated with river flow and saline intrusion. The second component, F2 (mass component MC2), results from (iii) the semi-diurnal SPM component generated by the combination of $U_1 \cos \omega t$ and $U_2 \cos (2\omega t - \theta)$ in (7.26) advected by (iv) the semi-diurnal current $U_1 \cos \omega t$. The first ($\cos \theta$) part of the resulting flux may be interpreted as representing the downstream ($U_2$ is negative from (7.25)) export of coarser (large $\alpha$) sediments. Equivalently, the second part ($\sin \theta$) represents upstream import of finer sediments (small $\alpha$). The third component, F3 (mass component MC3), is similar to the second, except that the semi-diurnal SPM component (v) arises from the combination of $U_0$ and $U_1 \cos \omega t$ in (7.26), i.e. MC3 in (7.29). Since $U_0 \sim U'_0 \sim U_2 \cos \theta$, this third flux component effectively increases the ($\cos \theta$) term in component (ii) by a factor of 3 (see Section 7.5).

Depth-integrated fluxes, corresponding to the conditions for Fig. 7.1(a), are shown (i) in Fig. 7.1(b) for a cyclical tidal numerical simulation and (ii) in Fig. 7.1(c) for the above analytical expansion (7.31). For these fluxes, agreement is not as close as for concentrations. However, the essential patterns and general magnitudes are sufficiently similar to warrant adoption of the analytical emulator for subsequent analyses of the characteristics and sensitivities of net sediment fluxes in tidal estuaries.

Figure 7.1(b) and (c) shows maximum upstream sediment fluxes exceeding $1000 \, t \, \mathrm{m}^{-1}$ width per year occurring for the finest sediment in the shallowest water. This value reduces for coarser material and in deeper water, reversing direction for $W_s > 0.002 \, \mathrm{m \, s}^{-1}$, with maximum downstream flux for $W_s \sim 0.005 \, \mathrm{m \, s}^{-1}$. By comparison, estimates of sediment deposition in the Mersey River over the last century involve net sediment fluxes of order of $(1000 \, t \, \mathrm{year}^{-1} \, \mathrm{m}^{-1})$ (Lane, 2004).

## 7.5 Component contributions to net sediment flux

Having demonstrated the validity of the analytical emulator in Section 7.4, we now apply the emulator to examine the component contributions to net sediment fluxes and, specifically, to identify conditions consistent with zero net fluxes, i.e. bathymetric stability.

### 7.5.1 Components of sediment flux: river flow, salinity and tidal current constituents

We note, from Section 7.2, that the components of residual velocities $U_R$ and $U_s$ associated with both river flow and salinity are typically two orders of magnitude less than $U_1$. Prandle (2003) calculated, for a random selection of 25 UK estuaries, an average value on extreme spring tides of $a = \varsigma/D = 2/3$. Then from (7.25), we estimate that the components of $U_0$ and $U_2$ related to varying estuarine cross section are typically as follows: (i) for $U_2$, comparable but less than $U_1$ and (ii) for $U_0$, an order of magnitude less than $U_1$.

For $\varsigma^* = 2$ m and $f = 0.0025$, Fig. 7.2(a) and (b) (Prandle, 2004c) shows residual sediment fluxes associated with (i) river flow $U_R = -0.01$ m s$^{-1}$ and (ii) salinity velocity $U_S$ from (7.10). River flow produces net downstream fluxes which increase with greater water depths and with finer sediments. Saline intrusion produces upstream fluxes increasing with greater depths and shows a maximum for $W_s \sim 0.002$ m s$^{-1}$. We see that the net fluxes associated with F1 in (7.31) are significantly less than for F2 and F3.

Figure 7.2(c) and (d) shows net fluxes associated with the the tidal coupling terms, F2 and F3. The relative magnitude of these coupling terms is directly related

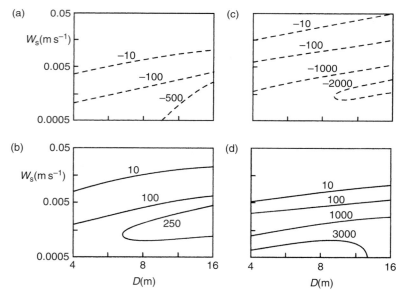

Fig. 7.2. Components of net depth-integrated sediment flux from (7.31).
(a) F1, river flow $U_R = -0.01$ m s$^{-1}$;
(b) F1, saline intrusion velocities, (7.10);
(c) cosine terms in components F2 and F3 of (7.31);
(d) sine terms in component F2.
Conditions, conventions and units as for Fig. 7.1.

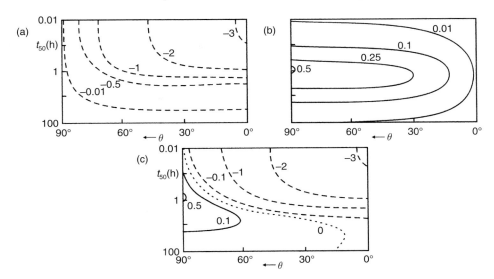

Fig. 7.3. Values of net import (+ve) and export terms in (7.31) as $f(\theta, t_{50})$. (a) Cosine terms, (b) sine term and (c) summation of (a) and (b).

to the phase difference $\theta$ between $M_2$ tidal elevation and current. The general nature of this balance between the 'cosine' and 'sine' terms is shown in Fig. 7.3 over a range of values of half-life in suspension, $t_{50}$, and $\theta$. Their contributions vary in a complex fashion given the counteracting sensitivities of $U_1$ and $\theta$ to changes in depth, (7.7) and (7.8). Likewise, their sensitivity to half-lives in suspension, $t_{50}$ (or $W_s$), is reflected in the dependence of the cosine and sine terms on $\alpha/(\alpha^2 + \omega^2)$ and $\omega/(\alpha^2 + \omega^2)$, respectively. Combining the sensitivities of these terms in (7.31) to both $\alpha$ and $\theta$, we anticipate that for the finest sediment in shallow water ($\theta \rightarrow 90°$), the $\sin \theta$ term will predominate, while for coarser sediments in deeper waters ($\theta \rightarrow 0°$), the $\cos \theta$ term will predominate.

From (7.31), assuming the term involving $U_1$ predominates in F1 and that $U_0 \sim U'_0 = -aU_1 \cos \theta$, then the ratio of F1 to F2 + F3 terms can be simplified to

$$\frac{0.5\, U_1^2\, U_{RS}/\alpha}{0.5\, a\, U_1^3\, (3\, a \cos \theta - \omega \sin \theta)/(\alpha^2 + \omega^2)} \tag{7.32}$$

(neglecting the parameters $Q$ and $P$ and the effect of vertical profiles in concentration); noting that $U_{RS} \sim -0.01 \text{ m s}^{-1}$ and letting $r = \omega/\alpha$, (7.32), simplifies to

$$\frac{-0.01}{-aU_1 \left( \dfrac{3 \cos \theta}{1 + r^2} - \dfrac{r \sin \theta}{1 + r^2} \right)}. \tag{7.33}$$

Figure 7.3(a) (Prandle, 2004c) represents $(-3\cos\theta)/(1+r^2)$, Fig. 7.3(b) represents $(r\sin\theta)/(1+r^2)$ and Fig. 7.3(c) represents the summation of these. We note that these terms balance in the region $1 < t_{50} < 10$ h when $t_{50} \sim 10(1-\theta/90)$ h. Further description of the significance of the parameter $r$ follows in Section 7.7.

### 7.5.2 Conditions for zero net flux of sediments

To convert from the generalised scaling results shown above to more specific estuarine conditions, we adopt the dynamical solutions for synchronous estuaries. This enables results to be presented in terms of the immediately familiar parameters, $\varsigma^*$ (tidal elevation amplitude) and $D$ (water depth), pertaining at any selected section of such estuaries.

Then for any combination of $(D, \varsigma^*)$, by scanning across a range of settling velocities, $W_s$, a balance between the predominant landward and the seaward tidal coupling terms in (7.33) can be found. Figure 7.4 (Prandle, 2004c) shows these 'zero flux' values for a (i) half-life $t_{50}$; (ii) particle diameter $d$, (7.21); (iii) fall velocity $W_s$, (7.22) and (iv) mean concentration, (7.29). The indicated range of values for particle diameter, $30 < d < 50\,\mu\text{m}$, is reasonably consistent with results described in Section 7.1.1 from Postma (1967), Uncles and Stephens (1989) and Hamblin (1989). Likewise, the range of values for settling velocity, $1 < W_s < 2.5\,\text{mm s}^{-1}$, is close to the value

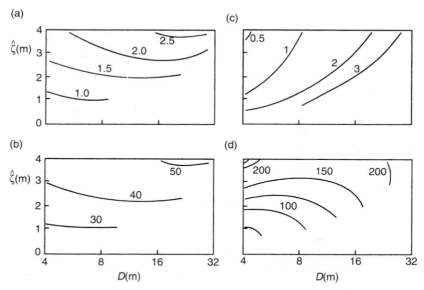

Fig. 7.4. Conditions for zero net sediment flux as $f(D, \varsigma^*)$, with $U_0 = -0.01\,\text{m s}^{-1}$ and $f = 0.0025$.
(a) Half-life $t_{50}$ (h),
(b) particle diameter ($\mu$m),
(c) fall velocity $W_s$ (mm s$^{-1}$) and
(d) mean sediment concentration (mg l$^{-1}$).

adopted for the Weser River by Lang *et al.* (1989). Moreover, there is recent evidence (Manning, 2004) from a number of estuaries of predominant settlement, via flocculation, at values of $W_s \sim 1 \, \mathrm{mm \, s}^{-1}$. The mean depth and tidally averaged suspended concentrations, $50 < C < 200 \, \mathrm{mg \, l}^{-1}$, are within the lower range of observed maximum concentrations, $10-10\,000 \, \mathrm{mg \, l}^{-1}$, found in *TM* (Uncles *et al.*, 2002). However, these estimates may be substantially increased by variations in the effective friction coefficient, here taken as $f = 0.0025$. An interesting feature of these results is the near constancy of the parameter $K_z / W_s D \sim 1.6$ throughout the range of $(D, \varsigma^*)$. It was noted earlier that maximum half-lives correspond to $K_z / W_s D = 2.5$, which coincides with $W_s \sim 1 \, \mathrm{mm \, s}^{-1}$. This indicates the coexistence of maximum concentrations with conditions of morphological stability.

### 7.5.3 Sensitivity to fall velocity, $W_s$

Figure 7.5(a) and (b) shows net sediment fluxes and mean suspended concentrations, calculated from (7.31) and (7.26), for the values $W_s = 1$ and $2 \, \mathrm{mm \, s}^{-1}$, suggested from Fig. 7.4(c) to be consistent with stable morphology. For the finer material, net fluxes are upstream for all but the lowest tides and deepest water. Moreover, these upstream fluxes increase roughly in proportion to $\varsigma^3$. Conversely,

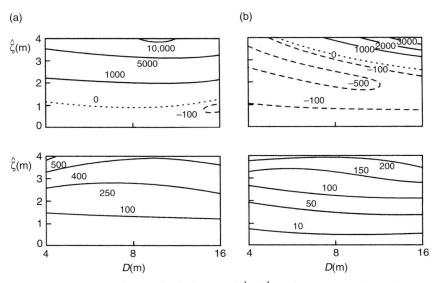

Fig. 7.5. Sediment fluxes (top) in $\mathrm{t \, year}^{-1} \mathrm{m}^{-1}$ and concentrations (bottom) in $\mathrm{mg \, l}^{-1}$ as $f(D, \varsigma^*)$. (a) Left, $W_S = 0.001 \, \mathrm{m \, s}^{-1}$ and (b) right, $W_S = 0.002 \, \mathrm{m \, s}^{-1}$. $D$ depth, $\varsigma$ tidal elevation amplitude.

for the coarser material, net fluxes are downstream for all but the largest tides and deepest water. These results clearly illustrate how both the direction and the magnitude of the net sediment fluxes can vary abruptly between spring and neap tides, from mouth to head, and for different sediment sizes.

### 7.5.4 Sensitivity to bed friction, f, and fall velocity, $W_s$

Corresponding calculations were made for the component contributions to these net fluxes. However, the scaling of the components shows less variation with either $\varsigma$ or $D$ than for variations in $W_s$ and the bed frictional coefficient $f$. To illustrate this,

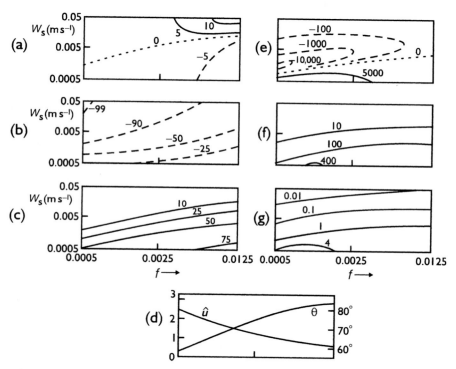

Fig. 7.6. Sensitivity of dynamics, sediment concentrations and net fluxes to fall velocity $W_S$ and friction factor $f$. Results for $\varsigma^* = 3$ m and $D = 8$ m.
(a)–(c). Percentage of net sediment flux in (7.31).
(a) F1,
(b) cosine flux term in F2 and F3,
(c) sine term in F2,
(d) tidal velocity $U^*$ and phase $\theta$,
(e) net flux F1 + F2 + F3 (t year$^{-1}$ m$^{-1}$),
(f) concentrations (mg l$^{-1}$) and
(g) half-life $t_{50}$ (h).

Fig. 7.6 (Prandle, 2004c) shows, for the case of $\varsigma = 3$ m and $D = 8$ m, these proportional components (in percentages) associated with (a) $U_{RS}$ (b) $\cos \theta$ and (c) $\sin \theta$ in (7.33) for $0.0005 < W_s < 0.05$ m s$^{-1}$ and $0.0005 < f < 0.0125$. These results show, consistent with Figs 7.4 and 7.5 and (7.28), that the cosine term predominates overall, producing a maximum (proportional) seaward flux for coarse sediments over the smoothest beds. Conversely, the sine term produces net upstream fluxes for the finest sediments over the roughest beds. Both of these results emphasise the reinforcement of axial sediment sorting noted in Section 7.5. By contrast, the role of the river and saline velocity components is generally negligible in all but the deepest waters. Moreover, while the salinity component predominates over the riverine component in transporting (near-bed) coarser material landward, the opposite occurs for (more uniformly distributed) finer sediments.

The significant sensitivity of both $U^*$ and $\theta$ to changes in bed friction is evident in Fig. 7.6(d) with consequent influences on concentrations and fluxes. Figure 7.6(e)–(g) indicates the related sensitivities of net flux, mean concentration and half-life in suspension.

## 7.6 Import or export of sediments?

This section summarises and interprets the application of earlier results for zero net sediment fluxes. Figure 7.7 presents a schematic of the dynamical and sedimentary components integrated into the analytical emulator used above to derive conditions corresponding to zero net flux of sediments, i.e. stable bathymetry.

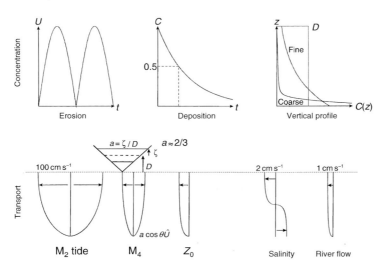

Fig. 7.7. Dynamical and sedimentary components integrated into the analytical emulator.

### 7.6.1 Sensitivity to half-life, $t_{50}$, and phase difference, $\theta$, between tidal elevation and current

The sensitivity of conditions corresponding to zero net flux of sediments are illustrated in Fig. 7.8 for (i) changes in sediment size, (ii) over the length of an estuary (indicated by changing depths) and (iii) over the spring–neap tidal cycle. Net transports of fine sediments are quantified in (7.33) and illustrated in Fig. 7.3(c). In Fig. 7.8 (Prandle *et al.*, 2005), loci of spring–neap variations for tidal elevation amplitudes $\varsigma = 1, 2, 3$ and 4 m are superimposed onto these results. These loci are for depths $D = 4$ and 16 m and for fall velocities $W_S = 0.01, 0.001$ and $0.0001\ \mathrm{m\,s^{-1}}$. The values for $\theta$ are from (7.7) adopting the assumption for bed friction coefficient $f = 0.001\ (d/10)^{1/2}$, where $d$ is particle diameter in micrometre and (7.21) relates $W_S$ to $d$. Corresponding values for $t_{50}$ are from (7.22).

Progressing from neap to spring tides increases (the absolute value of) $\theta$ and hence reduces the export or increases the import of sediments. This same trend is found in shallower depths, emphasising how estuaries can trap sediments in upstream sections. However, as more finer sediments are trapped, the effective value of $f$ decreases, resulting, (6.12), in a tendency to increase estuarine length. Hence, some equilibrium will prevail, governed by the balance between the type and

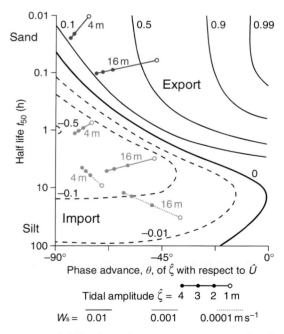

Fig. 7.8. Spring–neap variability in net import versus export of sediments (7.33), as a $f(t_{50}, \theta)$. Variability over elevation amplitude $\varsigma^* = 1, 2, 3$ and 4 m, fall velocities $W_S = 0.0001, 0.001$ and $0.01\ \mathrm{m\,s^{-1}}$ for depths $D = 4$ and 16 m.

the quantity of (marine) sediment supply, enhanced trapping of fine sediments and consequent more energetic dynamics attempting to increase depths.

From Fig. 5.2, for $K_z/W_sD < 2$, we can approximate $\alpha = W_s^2/K_z$, i.e. $r = \omega/\alpha = 3.5 \times 10^{-7} U^* D/W_s^2$. For typical values of $U^* \sim 1 \, \text{m s}^{-1}$ and $D \sim 10 \, \text{m}$, then $r < 1$ for $W_s > 2 \times 10^{-3} \, \text{m s}^{-1}$. Hence from (7.33), river flow and associated salinity intrusion (involving velocities of a few centimetres per second) will have little impact for sediments coarser than the above and as long as $(aU^*) \gg 0.01 \, \text{m s}^{-1}$, i.e. $a = \varsigma/D \gg 0.01$ (for the $M_2$ constituent).

Within these regions where tidal influences predominate, the ratio of sediment import (IM) to export (EX) is given by the denominator terms in brackets in (7.33):

$$\frac{\text{IM}}{\text{EX}} = \frac{r}{3}\tan\theta = 0.4 \left(\frac{f\,U^*}{W_s}\right)^2 \qquad (7.34)$$

on the basis of the above approximation for $\alpha$ and substituting for $\tan\theta$ from (7.7).

Zero net flux then corresponds to

$$W_s = 0.6\,f\,U^* \approx 0.0015\,U^* \qquad (7.35)$$

(for $f = 0.0025$). This estimate for $W_s$ is in close agreement with the values shown in Fig. 7.4(c).

The distribution for mean sediment concentration may be approximated by inserting $W_s$ from (7.35) in (7.29), neglecting $U_0$ and using the above values for $\alpha$, thus

$$\overline{C} = \gamma\rho f\,U^{*2}/D\alpha \approx \gamma\rho\;U^*(1 + a^2). \qquad (7.36)$$

Noting, as indicated in Section 7.2, that in shallow water, $U^*$ is proportional to $(\varsigma/f)^{1/2}D^{1/4}$, we see that for conditions of stable morphology, maximum concentrations will generally occur for large values of $a = \varsigma/D$ and small values of $f$. This sensitivity to $a$ is broadly substantiated in the concentrations shown in Fig. 7.5(a). However, the comparable sensitivities to $f$ shown for the specific conditions of Fig. 7.6 indicate a more complex relationship.

### *7.6.2 Sediment trapping and turbidity maxima*

From the above, we conclude that strongly tidal estuaries of the shape considered here will transport fine sediments upstream via non-linear tidal rectification terms. This process will be limited either by the absence of corresponding bed sediments for resuspension or by the counteracting influence of river flow close to the tidal limit.

Assuming that neap to spring tidal variations can be characterised as a single semi-diurnal constituent of varying amplitude, then springs will transport significantly more sediments; moreover, the maximum sediment size to be moved landward will be larger than at neaps (Fig. 7.8). To maintain morphological stability, these net spring and neap fluxes must balance. Hence, we anticipate a wide distribution and more continuous suspension of sediment sizes corresponding to this spring–neap varying demarcation, i.e. from Fig. 7.4(b) sediment diameters of typically 30–50 μm. Thus, in combination, these processes will produce trapping and 'sorting' of sediments with associated patterns varying over spring–neap and (close to the tidal limit) drought–flood cycles. However, this simplified approximation of the spring–neap cycle overlooks the role of accompanying constituents, in particular, the $MS_f$ component, which can significantly enhance the (apparent) $z_0$ constituent in this simplified examination of neap–spring variability.

The acute sensitivity of net fluxes to bed roughness shown in Fig. 7.6 also emphasises the feedback links to bed sediment distributions. These can occur over cycles of ebb and flood and spring and neaps tides as well as with gradual long-term changes and episodic extreme events.

The sorting mechanisms and axial accumulation of trapping over the cycles described will determine the size fraction and hence concentrations within the *TM* region. However, the location is likely to coincide with a maximum of $\varsigma/D$, modulated by the eventual predominance of river flow over tidal action towards the tidal limit.

### 7.6.3  Link to tidal energetics

Having shown that the balance of import to export of sediments depends directly on the phase lag, $\theta$, between tidal elevation and current, we note the direct corrspon-dence with net tidal energy dissipation which is proportional to $U^* \cos\theta$. Thus, we identify a relationship between the whole estuary tidal energy balance and the localised cross-sectional sediment flux balance. Such relationships have been sug-gested previously (Bagnold, 1963), and minimising net tidal energy dissipation is used as a stability condition in some morphological models.

## 7.7  Estuarine typologies

### 7.7.1  Bathymetry

Figures 7.9 and 7.10 (Prandle *et al.*, 2005) show observed lengths, $L$, and depths at the mouth, $D$, from 50 UK estuaries plotted as functions of $(Q, \varsigma^*)$, i.e. the mean river flow and the $M_2$ tidal amplitude at the mouth. The estuaries are restricted to those classified as either bar-built or coastal plain (Davidson and Buck, 1997). Corresponding theoretical values for $L$, (6.12), and $D$, (6.25) are shown for comparison. Overall, the observed values of depths and lengths are broadly consistent with the new dynamical theories.

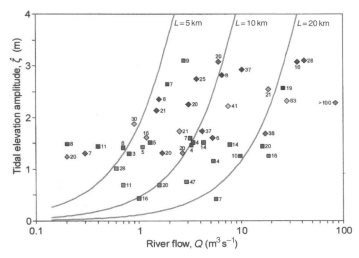

Fig. 7.9. Observed versus theoretical estuarine lengths, $L$(km), as $f(Q, \varsigma)$. Contours show theoretical values for $L$ (6.12). Observed data from estuaries shown in Fig. 6.12.

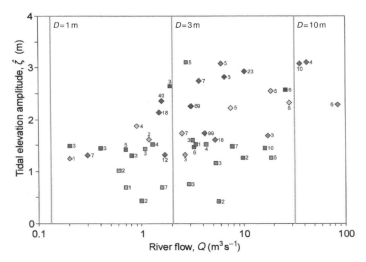

Fig. 7.10. Observed versus theoretical estuarine depths, $D$(m), (at the mouth) as $f(Q, \varsigma)$. Top axis shows theoretical values for $D$ (6.25), for side slope $\tan \alpha = 0.013$. Observed data from estuaries shown in Fig. 6.12.

The smaller depths in bar-built estuaries are clearly demonstrated. By identifying estuaries where depths diverge significantly from the theory, estimates can be made of the much larger flows existent in their post-glaciation formation. Regional discrepancies can also be used for inferring coasts with scarcity or plentiful supplies of sediment for infilling.

Assuming no in-fill but deepening via msl rise, Prandle *et al.* (2005) estimated the 'age', $N$ years, of estuaries, by least squares fitting of the expression $D_T(n) = D_O(n) - N\,S(n)$, where subscript T denotes theoretical and subscript O observed depths. $S(n)$ is the estimate, for each estuary, of the annual rate of relative msl rise, extracted for the estuaries shown in Fig. 6.12 from Shennan (1989). The msl trends vary from $-5$ to $2.0\,\text{mm}$ per year; over 10 000 years, this is equivalent to changes of $-5$ to $20\,\text{m}$ in estuarine depths. The resulting values were as follows: Rias, $N \sim 11\,000$ years; for coastal plain estuaries, $N \sim 15\,000$ years while for bar-built estuaries, $N \sim 100$ years. These results broadly confirm the morphological descriptions of their development, (Section 6.5.2).

### 7.7.2 Sediment regimes

This bathymetric typology is extended in Fig. 7.11 (Prandle *et al.*, 2005) to representations of the (depth and tidal) mean SPM concentrations (7.36) and the fall velocities consistent with zero net sediment flux (7.35). Additional observational data from three European estuaries are also included (Manning, 2004).

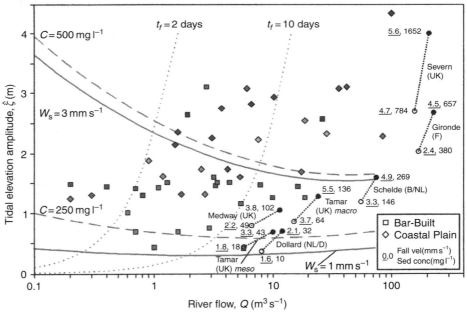

Fig. 7.11. 'Equilibrium' values of sediment concentrations and fall velocities. Estuarine flushing times. Observed versus equilibrium theory for: sediment concentrations $C$ (7.36) (dashed contours) and fall velocities $W_S$ (7.35) (full contours). Observed values for $W_S$ and $C$ in Dollard, Gironde, Medway, Schelde, Severn and Tamar (Manning, 2004). • spring, ○ neap tides. Flushing times from (7.37), dotted contour.

Note from (7.35) and (7.36) that the curves for fall velocity, $W_S$, and concentration, $C$, align directly with those for tidal current amplitude, $U^*$ (7.8). This typology illustrates why many estuaries show high levels of fine suspended sediments. The results for $W_s$ suggest a narrow range of fall velocities, typically between 1 and $3 \, \text{mm s}^{-1}$.

The results from Manning (2004) shown in Fig. 7.11 are representative of observed settlement of fine sediments in a wide range of European estuaries. These studies indicated that settling was primarily via the formation of micro- and macro-flocs, invariably close to the range suggested by the present theory. Likewise, prevailing observed suspended sediment concentrations are in good agreement with theoretical values.

Figure 7.11 also shows loci of representative flushing times, $T_F$, (4.60):

$$T_F = 0.5 \, \frac{L_I/2}{U_0} \qquad (7.37)$$

with $L_I$ from (7.11).

For river-borne dissolved or suspended sediments, the indicated values generally lie between 2 and 10 days (for residual current $U_0 = 1 \, \text{cm s}^{-1}$). These values are consistent with the range indicated from observations by Balls (1994) and Dyer (1997). Flushing times greater than the principal semi-diurnal tidal period provide valuable longer-term persistence of marine-derived nutrients, while flushing times less than the 15-day spring–neap cycle yield effective flushing of contaminants. Hence, there might be some ecological advantage to the bathymetric envelope defined by these two flushing times.

Assuming that fine marine sediment enters an estuary in almost continuous suspension (like salt) and none leaves, an estimate of minimum in-fill rates, $I_F$, can be obtained from (Prandle, 2004a):

$$I_F = \frac{\rho_s T_F}{0.69 \, C}, \qquad (7.38)$$

where $\rho_s$ is the density of the deposited sediment and $C$ is the mean suspended concentration from (7.36).

Prandle (2003) showed that for $U_0 \sim 0.01 \, \text{m s}^{-1}$, (7.38) indicates in-fill times ranging from 25 to 5000 years for depths from 5 to 30 m. Noting that 'capture rates' are typically only a few percent of entry rates (Lane and Prandle, 2006), we expect these minimum times to increase by one or two orders of magnitude.

## 7.8 Summary of results and guidelines for application

Synchronous estuary solutions for tidal dynamics and salinity intrusion derived in Chapter 6 are extended to include erosion, suspension and deposition of sediments. Integrating these processes into an 'analytical emulator' yields explicit expressions for cross-sectional fluxes of suspended sediments – enabling conditions of zero net flux (bathymetric stability) to be identified. Thus, it is shown how the exchange of sediments switches from export towards import as the ratio of tidal amplitude to depth increases and as sediment size decreases, providing quantitative explanations for trapping, sorting and *TM* in estuaries.

The leading questions are:

***What causes trapping, sorting and high concentrations of suspended sediments?***

***How does the balance of ebb and flood sediment fluxes adjust to maintain bathymetric stability?***

Suspended concentrations of fine sediments in tidal estuaries typically range from 100 to more than $1000 \, \text{mg} \, l^{-1}$, whereas concentrations in shelf seas are invariably less than $10 \, \text{mg} \, l^{-1}$. Moreover, observational and numerical modelling studies, Chapter 8, indicate that only a small fraction of the net tidal flux of sediments is permanently deposited. By introducing a 'synchronous estuary' assumption, Chapter 6 showed how estuarine bathymetries are determined by the tidal elevation amplitude, $\zeta^*$, and river flow, $Q$, alongside the bed friction coefficient $f$ (a proxy representation of the alluvium). Since this theory takes no account of the prevailing sediment regime, a paradigm reversal emerged – suggesting that sediment regimes are a consequence of, rather than a determinant for bathymetry.

To explain and explore this new paradigm and address the above questions, here we extend these 'synchronous' solutions for tidal dynamics and salinity intrusion to include sediment dynamics. (A synchronous estuary is one where axial surface gradients associated with phase changes greatly exceed those from amplitude variations.) Sediment erosion is reduced to its simplest format, i.e. proportional to velocity at the bed squared. Postma's (1967) description of delayed settlement in suspension is introduced by the adoption of exponential settling rates, with associated half-lives, $t_{50}$, based on a localised 'vertical dispersion-advective settling' model described in Chapter 5. This sediment module is combined with the dynamical solutions from Chapter 6 to form an 'analytical emulator', providing explicit expressions for concentrations and cross-sectional fluxes of sediment. The validity of these analytical expressions is assessed by comparison with detailed numerical model simulations. Figure 7.1 shows good agreement for both concentrations and net fluxes over a wide range of fall velocities, $W_s$, and water depths, $D$.

### 7.8.1 *Component contributions and sensitivity analyses*

Net contributions to sediment fluxes from residual currents associated with salinity intrusion, river flow and tidal non-linearities are shown in Fig. 7.2. Sediment fluxes were estimated by the product of these residual currents, $U_s + U_R + U_{M2}$ (as functions of depth) with related constituents of sediment concentration and their respective exponential depth variations. The resulting fluxes indicate complex patterns of net upstream or downstream movement acutely sensitive to $W_s$ and $f$. For estuaries with significant ratios of $\zeta^* : D$, i.e. substantial variations in cross-sectional area between high and low tidal levels, the contribution from the related generation of higher harmonic and residual current components, (7.25), far exceeds the other components.

### 7.8.2 Conditions for zero net flux of sediments (i.e. bathymetric stability)

Figure 7.3 illustrates how separate components from these tidal non-linearities, involving $\cos\theta$ and $\sin\theta$ (where $\theta$ is the phase difference between $\zeta^*$ and current $U^*$), determine the balance between import and export. From (7.32), combinations of $\theta$ and $t_{50}$ corresponding to zero net flux of sediments can be determined. Figure 7.4 shows 'zero flux' values for $t_{50}$, particle diameter $d$ and fall velocity $W_s$ alongside the related suspended sediment concentrations $C$ as functions of $\zeta^*$ and $D$.

To illustrate the acute sensitivity to $W_s$, Fig. 7.5 shows how these flux balances vary for a doubling of $W_s$ from 0.001 to 0.002 m s$^{-1}$. This sensitivity indicates how selective sorting occurs and explains how estuaries trap material in a size range 30–50 μm. It is shown that for zero net flux, $W_s \sim f\,U^*$. This latter relationship coincides with values of $K_Z/W_S D$ (the basic scaling parameter characterising suspended sediments), in the range 0.1–2.0 , i.e. close to conditions corresponding to maximum suspended sediment concentrations.

Figure 7.6 illustrates the sensitivity to the bed friction coefficient, $f$, of the components of net flux; suspended concentrations; tidal current amplitude and phase ($\varsigma^*$, $\theta$); and the half-life in suspension, $t_{50}$. Since net tidal energy dissipation is directly proportional to $\cos\theta$ , this sensitivity of $\theta$ to $f$ highlights a feedback link between the stability of estuarine-wide dynamics and both suspended and deposited sediments.

### 7.8.3 Neap to spring, coarse to fine, mouth to head variations in net import versus export

Figure 7.8 summarises these earlier results, illustrating how the balance between net import or export varies for depths from 4 to 16 m; fall velocities of 0.0001, 0.001 and 0.01 m s$^{-1}$ and tidal amplitudes from 1 to 4 m (representative of neap to spring variations). This indicates how in proceeding upstream from deep to shallow water, the balance between import of fine sediments and export of coarser ones becomes finer, i.e. selective 'sorting' and trapping. Likewise, more imports, extending to a coarser fraction, occur on spring than on neap tides.

Overall, the emulator provides new insights into the balance between tides, river flow and bathymetry and on the relationship of these with the prevailing sediment regime. The conditions derived for maintaining stable bathymetry extend earlier concepts of flood- and ebb-dominated regimes, with sediment import and export shown to vary axially, with sediment type, and over the spring to neap tidal cycle. These stable conditions are shown to correspond both with conditions for maximum SPM concentrations and with observations of the predominant settling rates in many estuaries.

### 7.8.4  Typological frameworks

These developments enable the bathymetric frameworks, developed for dynamics and saline intrusion in synchronous estuaries in Chapter 6, to be extended to indicate the nature of the sedimentary regimes consistent with bathymetric stability. Comparisons of these extended typological frameworks against observational data are shown in Figs 7.9–7.11.

By relating the difference between observed and theoretical depths to the years of (relative) sea level change since an estuary was formed, representative ages for Rias, Coastal Plain and Bar-Built estuaries are calculated.

The characteristics of dynamical and bathymetric estuarine parameters for 'mixed' estuaries, listed as (a) to (i) in Section 6.6, can be further extended to include the following sedimentary parameters:

$$(j) \text{ Suspended concentration } C \propto f U^* \qquad (7.39)$$

$$(k) \text{ Equilibrium fall velocity } W_s \propto f U^* \qquad (7.40)$$

Overall, the fit found between observed and theoretical bathymetries substantiates the earlier paradigm reversal (Chapter 6) that prevailing sediment dynamics are a consequence of selective sorting and trapping by the existing dynamics and bathymetry. It is then postulated that associated bathymetric evolution will be determined by changes in tidal amplitude, river flow and bed roughness. The rate of change will be modulated, or even superseded, by the supply of sediments and regional changes in relative sea level.

### References

Aubrey, D.G., 1986. Hydrodynamic controls on sediment transport in well-mixed bays and estuaries. In: Van de Kreeke, J. (ed.), *Physics of Shallow Estuaries and Bays*. Springer-Verlag, New York, pp. 245–285.

Bagnold, P.A., 1963. Mechanics of marine sedimentation. In: Hill, M.N. (ed.), *The Sea: Ideas and Observations on Progress in the Study of the Seas*, Vol. 3, *The Earth Beneath the Sea: History*. John Wiley, Hoboken, NJ, pp. 507–582.

Balls, P.W., 1994. Nutrient inputs to estuaries from nine Scottish East Coast Rivers: Influence of estuarine processes on inputs to the North Sea. *Estuarine, Coastal and Shelf Science*, **39**, 329–352.

Davidson, N.C. and Buck, A.L., 1997. *An Inventory of UK Estuaries. Vol. 1. Introduction and Methodology*. Joint Nature Conservatory Communication, Peterborough, UK.

Dronkers, J. and Van de Kreek, J., 1986. Experimental determination of salt intrusion mechanisms in the Volkerak estuary, Netherlands. *Journal Sea Research*, **20** (1), 1–19.

Dyer, K.R., 1997. *Estuaries: A Physical Introduction*, 2nd edn. John Wiley, Hoboken, NJ.

Dyer, K. and Evans, E.M., 1989. Dynamics of turbidity maximum in a homogeneous tidal channel. *Journal of Coastal Research*, **5**, 23–30.

Festa, J.F. and Hansen, D.V.C., 1978. Turbidity maxima in partially mixed estuaries: a two-dimensional numerical model. *Estuarine, Coastal and Marine Science*, **7**, 347–359.

Friedrichs, C.T, Armbrust, B.D., and de Swart, H.E., 1998. Hydrodynamics and equilibrium sediment dynamics of shallow, funnel-shaped tidal estuaries. In: Dronkers, J. and Schaffers, M. (eds), *Physics of Estuaries and Coastal Seas*, A.A. Balkema, Brookfield, Vt. pp. 315–328.

Grabemann, I. and Krause, G., 1989. Transport processes of suspended matter derived from time series in a tidal estuary. *Journal of Geophysical Research*, **94**, 14373–14380.

Hamblin, P.F., 1989. Observations and model of sediment transport near the turbidity maximum of the upper Saint Lawrence estuary. *Journal of Geophysical Research*, **94**, 14419–14428.

Jay, D.A. and Kukulka, T., 2003. Revising the paradigm of tidal analysis: The uses of non-stationary data. *Ocean Dynamics*, **53**, 110–123.

Jay, D.A. and Musiak, J.D., 1994. Particle trapping in estuarine turbidity maxima. *Journal of Geophyical Research*, **99**, 20446–20461.

Lane, A., 2004. Morphological evolution in the Mersey estuary, UK 1906–1997: Causes and effects. *Estuarine Coastal and Shelf Science*, **59**, 249–263.

Lane, A. and Prandle, D., 2006. Random-walk particle modelling for estimating bathymetric evolution of an estuary. *Estuarine, Coastal and Shelf Science*, **68** (1–2), 175–187.

Lang, G., Schubert, R., Markofsky, M., Fanger, H-U., Grabemann, I., Krasemann, H.L., Neumann, L.J.R., and Riethmuller, R., 1989. Data interpretation and numerical modeling of the mud and suspended sediment experiment 1985. *Journal of Geophysical Research*, **94**, 14381–14393.

Manning, A.J., 2004. Observations of the properties of flocculated cohesive sediments in three western European estuaries. In Sediment Transport in European Estuaries. *Journal of Coastal Research*, **SI 41**, 70–81.

Markofsky, M., Lang, G., and Schubert, R., 1986. Suspended sediment transport in rivers and estuaries. In: Van de Kreeke, J. (ed.) *Physics of Shallow Estuaries and Bays*. Springer-Verlag, New York, pp. 210–227.

Postma, H., 1967. Sediment transport and sedimentation in the estuarine environment. In: Lauff, G.H. (ed.), *Estuaries*, Publication No. 83. American Association for the Advancement of Science, Washington, DC, pp. 158–179.

Prandle, D., 1985. On salinity regimes and the vertical structure of residual flows in narrow tidal estuaries. *Estuarine, Coastal and Shelf Science*, **20**, 615–633.

Prandle, D., 1997a. The dynamics of suspended sediments in tidal waters. *Journal of Coastal Research*, **25**, 75–86.

Prandle. D., 1997b. Tidal characteristics of suspended sediment concentrations. *Journal of Hydraulic Engineering*, **123** (4), 341–350.

Prandle, D., 2003. Relationships between tidal dynamics and bathymetry in strongly convergent estuaries. *Journal of Physical Oceanography*, **33** (12), 2738–2750.

Prandle, D., 2004a. How tides and river flows determine estuarine bathymetries. *Progress in Oceanography*, **61**, 1–26.

Prandle, D., 2004b. Saline intrusion in partially mixed estuaries. *Estuarine, Coastal and Shelf Science*, **59** (3), 385–397.

Prandle, D., 2004c. Sediment trapping, turbidity maxima and bathymetric stability in macro-tidal estuaries. *Journal of Geophysical Research*, **109** (C08001), 13pp.

Prandle, D., Lane, A. and Manning, A.J., 2005. Estuaries are not so unique. *Geophysical Research Letters*, **32** (23), L23614.

Prandle, D., Lane, A., and Wolf, J., 2001. Holderness coastal erosion – Offshore movement by tides and waves. In: Huntley, D.A., Leeks, G.J.J., and Walling, D.E. (ed), *Land-Ocean Interaction, Measuring and Modelling Fluxes from River Basins to Coastal Seas*. IWA publishing, London, pp. 209–240.

Shennan, I., 1989. Holocene crustal movements and sea-level changes in Great Britain. *Journal of Quaternary Science*, **4** (1), 77–89.

Uncles, R.J. and Stephens, J.A., 1989. Distributions of suspended sediment at high water in a macrotidal estuary. *Journal of Geophysical Research*, **94**, 14395–14405.

Uncles, R.J., Stephens, J.A., and Smith, R.E., 2002. The dependence of estuarine turbidity on tidal intrusion length, tidal range and residence times. *Continental Shelf Research*, **22**, 1835–1856.

# 8

# Strategies for sustainability

## 8.1 Introduction

Rising sea levels and enhanced storminess, resulting from Global Climate Change (GCC), pose serious concerns about the viability of estuaries worldwide. This book has shown how related impacts are manifested via evolving interactions between tidal dynamics, salinity intrusion, sedimentation and morphology.

Drawing on a case study of the Mersey Estuary, Section 8.2 assesses the capabilities of a fine-resolution 3D model against the perspective of a 100-year record of changes in tides, sediments and estuarine bathymetries. Both historical and recent observations are used alongside the Theoretical Frameworks, described in preceding chapters, to interpret ensemble simulations of parameter sensitivities and so reduce levels of uncertainty surrounding future forecasts.

There is an urgent need to develop models that can indicate the possible nature, extent and rate of morphological changes. With accurate information on bathymetry and surficial sediment distribution, 'Bottom-Up' numerical models (i.e. solving momentum and continuity equations as in the Mersey study) can accurately reproduce water levels and currents. However, simulation of sediment regimes involves net fluxes generally determined by non-linear coupling between flow and sediment suspension, i.e. processes over much wider spectral scales. Future forecasts must encapsulate a broad spectrum of changing conditions, consequently the range of possible future morphologies widens sharply. While 'Bottom-Up' numerical models can be used for sensitivity analyses to identify areas susceptible to erosion or deposition, longer-term extrapolations become increasingly chaotic. As an alternative to 'Bottom-Up' models for forecasting bathymetric evolution, geomorphologists use 'Top-Down', 'rule-based' models (Pethick, 1984). A number of 'stability criteria' are used, based generally on fitting observed bathymetry to simplified dynamical criteria. Since this fitting often extends over millennia, these 'Top-Down' approaches provide valuable long-term perspectives.

In Section 8.3, the explicit formulae and Theoretical Frameworks are used to make future predictions regarding likely impacts from GCC. Quantitative estimates are made of possible changes in estuarine bathymetries up to 2100. Attendant consequences for tides, storms, salinity intrusion and sediment regimes are discussed.

Section 8.4 indicates strategies for long-term management of estuaries, with sections on modelling, observations, monitoring and forecasting. Appendix 8A emphasises the potential of operational modelling and monitoring in global-scale efforts to address GCC.

## 8.2 Model study of the Mersey Estuary

This case study illustrates how the formulation and validation of a detailed numerical model can utilise observational data ranging from process 'measurements', extended 'observations' to permanent 'monitoring'. Likewise, it shows how the Theoretical Frameworks, described in earlier chapters, are used to interpret ensemble sensitivity simulations.

### 8.2.1 Tidal dynamics, sediment regime and bathymetric evolution of the Mersey

Tidal ranges in the Mersey vary from 4 to 10 m over the extremes from neap to spring. The estuary has been widely studied because of its vital role in shipping. The 'Narrows' at the mouth of the 45-km long estuary is approximately 1.5-km wide with a mean depth (below chart datum) of 15 m (Fig. 8.1; Lane and Prandle, 2006). Tidal currents through this section can exceed $2 \, \text{m s}^{-1}$. Further upstream in the inner estuary basin, the width can be as much as 5 km, and extensive areas are exposed at low water. Freshwater flow into the estuary, $Q$, varies from 25 to $300 \, \text{m}^3 \, \text{s}^{-1}$ with a mean 'flow ratio' ($Q \times 12.42$ h/volume between high and low water) of approximately 0.01. Flow ratios of less than 0.1 usually indicate well-mixed conditions (4.66), though in certain sections during part of the tidal cycle, the Mersey is only partially mixed.

### Suspended sediments and net deposition

Figure 8.2 ( Lane and Prandle, 2006) shows observed suspended sediment time series from locations in the Narrows recorded in 1986 and 1992; Table 8.1 summarises these results. The 1986 observations included five simultaneous moorings across the Narrows, providing estimates of net spring and neap tidal fluxes of sediments. Prandle *et al.* (1990) analysed four sets of observations of SPM indicating tidally averaged cross-sectional mean concentrations varying as a function of tidal amplitudes, $\varsigma^*$, as follows: $32 \, \text{mg l}^{-1}$ for $\varsigma^* = 2.6 \, \text{m}$, $100 \, \text{mg l}^{-1}$ for $\varsigma^* = 3.1 \, \text{m}$, $200 \, \text{mg l}^{-1}$ for $\varsigma^* = 3.6 \, \text{m}$ and $213 \, \text{mg l}^{-1}$ for $\varsigma^* = 4.0 \, \text{m}$. These values correspond to a tidal flux

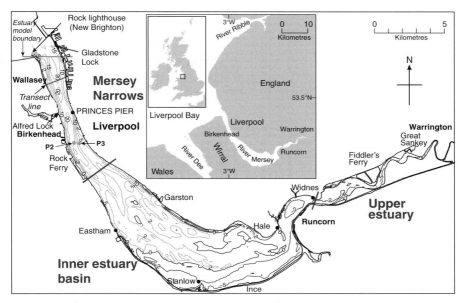

Fig. 8.1. Liverpool Bay and the Mersey Estuary location map. The 1992 transect line corresponds to positions P2, P3; tide gauges are marked with dots. Depths (1997 bathymetry) are in metres below Ordnance Datum Newlyn (ODN). Chart datum is approximately the lowest astronomical tide level and is 4.93 m below ODN.

(on ebb or flood) of 40 000 t on a mean tide, reducing to as little as 2500 t at neap and increasing by up to 200 000 t on springs – in reasonable agreement with earlier estimates from a hydraulic model study by Price and Kendrick (1963).

Hutchinson and Prandle (1994) used contaminant sequences in sediment cores (analogous to tree-ringing) to estimate net accretion rates in the adjacent and the similarly sized Dee Estuary. These amounted to 0.3 Mt a$^{-1}$ between 1970 and 1990 and 0.6 Mt a$^{-1}$ between 1950 and 1970. Using *in situ* bottle samples, Hill *et al.* (2003) derived settling velocities, $W_S$, of 0.0035 m s$^{-1}$ for spring tidal conditions and 0.008 m s$^{-1}$ for neaps. Noting that particle diameter $d$ (μm) $\sim 1000\, W_S^{1/2}$ (m s$^{-1}$), these correspond to $d = 59$ and 89 μm, respectively.

### Bathymetry

Data were available from surveys carried out by the Mersey Docks and Harbour Company in 1906, 1936, 1956, 1977 and 1997. Differences in net volume within the Narrows are of the order of a few percent from one data set to the next. The largest changes appear in the inter-tidal regions of the inner estuary basin, particularly from Hale and Stanlow to Runcorn where the low water channel positions change readily, and volume differences between successive surveys exceed 10%. The overall

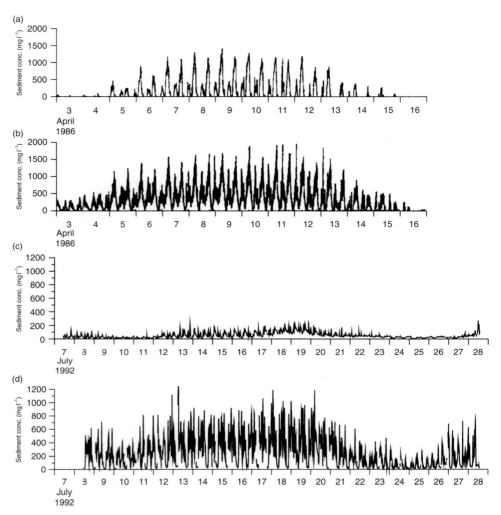

Fig. 8.2. Observed sediment concentrations in the Narrows. (a) surface, (b) mid-depth in 1986, (c) 280 m from Wirral shore and (d) 290 m from Liverpool shore in 1992.

pattern is a decrease in estuary volume of about $60\,\mathrm{mm}^3$ or 8% between 1906 and 1977 despite sea level rise averaging $1.23\,\mathrm{mm}$ per year during the past century (Woodworth *et al.*, 1999). After this period, there is a small increase of $10\,\mathrm{mm}^3$.

### 8.2.2 Modelling approach

Here we illustrate the capabilities and limitations of a 3D Eulerian hydrody-namic model coupled with a Lagrangian sediment module (Lane and Prandle,

Table 8.1 *Suspended sediment concentrations, net deposition and net tidal fluxes*

| | Suspended sediment concentration ($mg\,l^{-1}$) at position (2) | | | Net sediment deposited ($10^3\,t\,a^{-1}$) | Net tidal flux ($10^3\,m^3\,s^{-1}$) | | |
|---|---|---|---|---|---|---|---|
| | Mean | Max | Min | | Spring | Neap | |
| Sediment settling velocity $W_s$ ($ms^{-1}$) | | | | | | | |
| 0.005 | *25* | *67* | *0* | *1800* | *46.5* | *2.3* | *Lagrangian model* |
| 0.0005 | *213* | *442* | *0* | *4900* | *306.0* | *8.8* | *Lagrangian model* |
| Observed in 1986 | 300 | 1100 | 0 | | 200.0[b] | 60.0[b] | Observed (surface) |
| | 500 | 1500 | 0 | | | | Observed (mid-depth) |
| 1992 (1) | 53 | 115[d] | 0 | | | | |
| (2) | 250 | 1500[d] | 0 | 2300[a,c] | | | Bathymetric records |

[a]Lane (2004).
[b]Prandle *et al.* (1990).
[c]Thomas *et al.* (2002).
[d]90% of sediments have concentrations less than this value.

*Notes*: Observed Values: Transect line P2–P3 (Fig. 8.1): (1) 280 m from Wirral; (2) 290 m from Liverpool shore.
*Source*: Lane and Prandle, 2006.

2006) to quantify impacts on the estuarine sediment regime and indicate the rate and nature of bathymetric evolution. Particular emphasis is on quantifying the variations in sediment concentrations and fluxes in sensitivity tests of bed roughness, eddy viscosity, sediment supply (particle sizes 10–100 µm), salinity intrusion and 2D versus 3D formulations of the hydrodynamic model. The model was not intended to reproduce bed-load transport associated with coarser sediments.

Recognising the limited capabilities to monitor the often extremely heterogeneous SPM, a wide range of observational data were used for assessing model performance. These include suspended concentrations (axial profiles of mean and '90th percentile'), tidal and residual fluxes at cross sections, estuary-wide net suspension and deposition on spring and neap tides, surficial sediment distributions and sequences of bathymetric evolution.

The Eulerian hydrodynamic model provides velocities, elevations and diffusivity coefficients for the Lagrangian 'random-walk' particle model in which up to a million particles represent the sediment movements. It includes a wetting-and-drying scheme to account for the extensive inter-tidal areas. Forcing involved specifying tidal elevation constituents at the seaward limit in the Mersey Narrows, and river flow at the head. The model uses a 120-m rectangular grid horizontally and a 10-level sigma-coordinate scheme in the vertical.

Calibration of the model (Lane, 2004) involved simulating effects of 'perturbations', i.e. varying the msl, bed friction coefficients, vertical eddy viscosity and the river flow. The optimum combination to minimise differences between observed and modelled tidal elevation constituents was then determined. The model was most sensitive to changes in bathymetries and bed friction coefficients, particularly in the inner basin. River flow only has an appreciable effect for discharges significantly higher than those usually encountered.

### 8.2.3  *Lagrangian, random-walk particle module for non-cohesive sediment*

Random-walk particle models replicate solutions of the Eulerian advection–diffusion equation by calculating, for successive time steps $\Delta t$, the height above the bed, $Z$, and horizontal location of each particle following

(1) a vertical advective movement – $W_S \Delta t$ (downwards),
(2) a diffusive displacement $l$ (up or down),
(3) horizontal advection.

The displacement length $l = \sqrt{(2K_z\Delta t)}$ (Fischer *et al.*, 1979), with the vertical eddy viscosity coefficient $K_z$ approximated by $fU^*D$, where $f$ is the bed friction coefficient, $U^*$ the tidal current amplitude and $D$ depth. Contacts with the surface and bed during this diffusion step are reflected elastically. Deposition occurs when the particle reaches the bed during a discrete advective settlement step – $W_S$ $\Delta t$. For any grid square containing deposited particles, new particles are released into suspension by time-integration of the erosion potential.

A simple algorithm for the erosion source was adopted:

$$ER = \gamma \rho f U^P, \tag{8.1}$$

where $\rho$ is water density and a value of $P = 2$ was assumed. Having specified $P$, all subsequent calculations of concentration, flux and sedimentation rates are linearly proportional to the coefficient $\gamma$. A value of $\gamma = 0.0001 \text{ m s}^{-1}$ was found to produce suspended sediment concentrations comparable with those in Fig. 8.2. The corresponding values of tidal and residual cross-sectional fluxes were also in reasonable agreement with observed values shown in Table 8.1.

### 8.2.4 Simulations for fall velocities, $W_S = 0.005$ and $0.0005 \text{ m s}^{-1}$

Figure 8.3 (Lane and Prandle, 2006) shows cross-sectional mean suspended sediment concentrations, at successive locations landwards from the mouth, over two spring–neap cycles commencing from the initial introduction of sediments. The examples chosen are for sediment fall velocities, $W_S$ of $0.005 \text{ m s}^{-1}$ (coarse sediment $d = 70 \,\mu\text{m}$, black lines) and $0.0005 \text{ m s}^{-1}$ (finer sediment, $d = 22 \,\mu\text{m}$, grey lines), respectively.

Starting with no sediment in the estuary, all particles are introduced at the seaward boundary of the model using the erosion formula (8.1). An unlimited supply is assumed together with zero axial concentration gradient ($\partial C/\partial X = 0$) for inflow conditions. To reflect the effect of changing distributions of surficial sediments on the bed friction coefficient, this was specified as $f = 0.0158 \, W_S^{\frac{1}{4}}$.

For $W_S = 0.0005 \text{ m s}^{-1}$ (grey lines), the suspended sediment time series change from predominantly semi-diurnal (linked to advection) at the mouth to quarter-diurnal (linked to localised resuspension) further upstream. Even close to the mouth, a significant quarter-diurnal component is generated at spring tides. Close to the mouth, peak concentrations occur some three tidal cycles after maximum spring tides, while further upstream this lag extends up to seven cycles.

For the coarser sediment, $W_S = 0.005 \text{ m s}^{-1}$ (black lines), Fig. 8.3 shows much reduced concentrations largely confined to the seaward region, although the slower 'adjustment' rate suggests that a longer simulation is required to introduce the coarser sediments further upstream. The time series is predominantly quarter-diurnal and peak concentrations coincide with peak tides; the sediments have a much shorter half-life in suspension as described in Section 5.5.

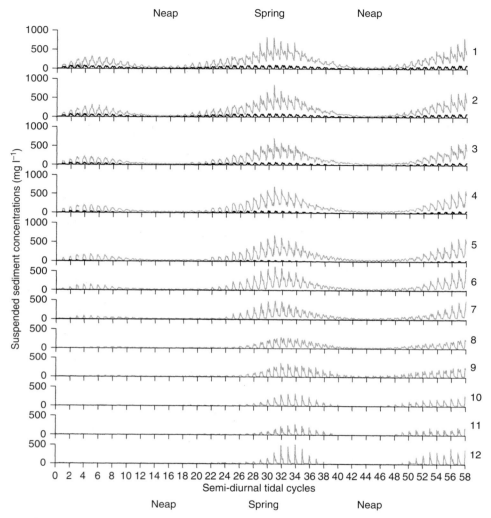

Fig. 8.3. Suspended sediment concentrations at 12 positions along the Mersey 1 the mouth, 12 the head; grey lines settling velocity $W_S = 0.0005$ m s$^{-1}$; black lines $W_S = 0.005$ m s$^{-1}$.

Figure 8.4(a) (Lane and Prandle, 2006) shows corresponding time series of cumulative inflow and outflow of sediments across the mouth of the estuary model. Differences between inflow and outflow, in Fig. 8.4(b), indicate net suspension (high frequency) and net deposition (low frequency). For $W_S = 0.0005$ m s$^{-1}$, the mean tidal exchange of sediments is around 110 000 t per tide, of which approximately 6% is retained amounting to 7000 t per tide. For $W_S = 0.005$ m s$^{-1}$, the mean exchange is 22 000 tonnes of which approximately 12% is retained or about 3000 t per tide.

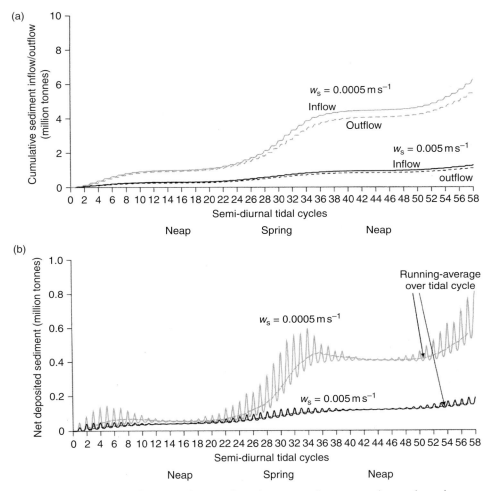

Fig. 8.4. Sediment fluxes at the mouth and net estuarine suspension and erosion.
(a) Cumulative inflow and outflow at the mouth of the Mersey. (b) Net suspension
(high frequency) and deposition (low frequency).

### 8.2.5 Sensitivity to sediment size

Full details of the sensitivity tests are shown by Lane and Prandle (2006), these are
summarised in Tables 8.2 and 8.3. For a more extensive quantitative evaluation of
the model, single neap–spring tidal cycle simulations were used. Results are summarised in Table 8.2 for particle diameters $d$ from 10 to 100 μm.

The model reveals that mean suspended sediment concentrations vary approximately with $d^{-2}$. Equation (7.29a) indicates variability ranging from $d^0$ to $d^{-4}$ for
finer to coarser sediments. The extent of landward intrusion increases progressively
for finer sediments. A minimum capture rate of 2.8% occurs for $d = 30$ μm with

Table 8.2 *Sensitivity of modelled sediments to particle diameters d from 10 to 100 μm for (a) maximum (90th percentile) concentrations; (b) mean concentrations; (c) estuarine-wide mean suspended and net deposited sediments on neap and spring tides*

(a)

90th percentile suspended sediment concentrations (mg l$^{-1}$) at 2 km intervals upstream from the mouth

| $d$(μm) | | | | | | | | | | | | | | | | | |
|---|---|---|---|---|---|---|---|---|---|---|---|---|---|---|---|---|---|
| 10 | 4186 | 3896 | 3804 | 3366 | 3274 | 3167 | 2471 | 1525 | 1052 | 714 | 560 | 470 | 500 | 348 | 309 | 217 | 162 |
| 20 | 369 | 344 | 325 | 297 | 258 | 240 | 218 | 183 | 140 | 84 | 52 | 37 | 36 | 20 | 19 | 20 | 12 |
| 30 | 169 | 148 | 133 | 101 | 89 | 67 | 47 | 29 | 12 | 6 | 3 | 2 | 2 | — | — | — | — |
| 40 | 140 | 118 | 103 | 75 | 64 | 45 | 26 | 12 | 5 | 3 | 2 | 1 | 1 | — | — | — | — |
| 50 | 108 | 95 | 83 | 62 | 53 | 35 | 16 | 7 | 4 | 3 | 2 | 1 | 1 | — | — | — | — |
| 60 | 83 | 72 | 65 | 47 | 39 | 22 | 8 | 3 | 1 | — | — | — | — | — | — | — | — |
| 70 | 63 | 56 | 50 | 38 | 29 | 16 | 5 | 1 | — | — | — | — | — | — | — | — | — |
| 80 | 50 | 45 | 40 | 30 | 22 | 11 | 2 | — | — | — | — | — | — | — | — | — | — |
| 90 | 39 | 35 | 29 | 22 | 16 | 7 | — | — | — | — | — | — | — | — | — | — | — |
| 100 | 31 | 29 | 25 | 17 | 12 | 4 | — | — | — | — | — | — | — | — | — | — | — |

(b)

Mean suspended sediment concentrations (mg l$^{-1}$) at 2 km intervals upstream from the mouth

| $d$(μm) | | | | | | | | | | | | | | | | | |
|---|---|---|---|---|---|---|---|---|---|---|---|---|---|---|---|---|---|
| 10 | 1645 | 1465 | 1423 | 1183 | 1118 | 1049 | 875 | 669 | 562 | 418 | 252 | 177 | 173 | 119 | 106 | 81 | 92 |
| 20 | 176 | 157 | 151 | 128 | 120 | 108 | 93 | 75 | 60 | 43 | 27 | 17 | 15 | 8 | 7 | 5 | 5 |
| 30 | 68 | 56 | 47 | 34 | 27 | 20 | 14 | 8 | 4 | 2 | 1 | — | — | — | — | — | — |
| 40 | 51 | 41 | 33 | 22 | 17 | 11 | 6 | 3 | 1 | 1 | — | — | — | — | — | — | — |
| 50 | 41 | 33 | 27 | 19 | 14 | 8 | 4 | 2 | 1 | — | — | — | — | — | — | — | — |
| 60 | 32 | 26 | 20 | 14 | 10 | 5 | 2 | — | — | — | — | — | — | — | — | — | — |
| 70 | 23 | 19 | 15 | 10 | 7 | 3 | — | — | — | — | — | — | — | — | — | — | — |
| 80 | 18 | 14 | 11 | 7 | 5 | 2 | — | — | — | — | — | — | — | — | — | — | — |
| 90 | 13 | 10 | 8 | 5 | 3 | 1 | — | — | — | — | — | — | — | — | — | — | — |
| 100 | 10 | 8 | 6 | 3 | 2 | — | — | — | — | — | — | — | — | — | — | — | — |

(c)

| d (μm) | Suspended | | Deposited | | Exchange | | Deposited Per year | % deposit per Exchange | Average suspended |
|---|---|---|---|---|---|---|---|---|---|
| | Neap | Spring | Neap | Spring | Neap | Spring | | | |
| 10 | 156.21 | 1531.21 | 28.40 | 420.84 | 57.53 | 2508.81 | 66400 | 10.5 | 640.06 |
| 20 | 2.15 | 206.55 | -0.99 | 61.07 | 9.80 | 490.07 | 5000 | 5.9 | 75.75 |
| 30 | 3.97 | 34.77 | -0.67 | 4.93 | 5.86 | 106.96 | 1000 | 2.8 | 15.03 |
| 40 | 2.31 | 22.62 | -0.54 | 5.06 | 5.51 | 87.48 | 1200 | 4.4 | 9.46 |
| 50 | 1.77 | 18.07 | -0.05 | 6.12 | 4.14 | 71.20 | 1400 | 6.2 | 7.47 |
| 60 | 1.13 | 12.97 | -0.04 | 7.15 | 3.21 | 58.85 | 2000 | 10.5 | 5.19 |
| 70 | 0.79 | 9.28 | 0.56 | 6.49 | 2.93 | 43.83 | 1800 | 12.3 | 3.69 |
| 80 | 0.58 | 7.36 | 0.23 | 7.64 | 2.02 | 36.91 | 1700 | 14.0 | 2.87 |
| 90 | 0.41 | 5.48 | 0.14 | 7.54 | 1.70 | 30.55 | 1600 | 16.9 | 2.11 |
| 100 | 0.26 | 4.48 | 0.10 | 5.62 | 1.15 | 25.09 | 1400 | 18.6 | 1.64 |

*Notes:* Units: $10^3$ tonnes.
*Source:* Lane and Prandle, 2006.

Table 8.3   *Sensitivity of modelled sediments, R1–R8, for* $W_S = 0.0005 \, m \, s^{-1}$ *(d = 22 μ)*

(a)

Mean suspended sediment concentrations ($mg \, l^{-1}$) at 2 km intervals upstream from the mouth

| Run | | | | | | | | | | | | | | | | | |
|-----|---|---|---|---|---|---|---|---|---|---|---|---|---|---|---|---|---|
| (1) | 127 | 109 | 98 | 77 | 67 | 58 | 47 | 34 | 25 | 16 | 10 | 6 | 4 | 2 | 2 | 1 | 1 |
| (2) | 333 | 304 | 301 | 263 | 253 | 234 | 205 | 171 | 141 | 106 | 71 | 50 | 49 | 35 | 29 | 23 | 25 |
| (3) | 132 | 117 | 112 | 95 | 89 | 79 | 69 | 58 | 48 | 36 | 26 | 17 | 17 | 8 | 7 | 5 | 5 |
| (4) | 127 | 112 | 104 | 84 | 76 | 69 | 56 | 43 | 32 | 21 | 12 | 7 | 5 | 2 | 2 | 1 | 1 |
| (5) | 48 | 42 | 40 | 33 | 31 | 28 | 22 | 17 | 14 | 10 | 7 | 5 | 4 | 1 | 1 | 1 | 1 |
| (6) | 196 | 171 | 155 | 127 | 116 | 105 | 86 | 60 | 42 | 24 | 12 | 7 | 5 | 2 | 1 | 1 | 1 |
| (7) | 73 | 65 | 60 | 49 | 44 | 38 | 30 | 23 | 16 | 10 | 5 | 2 | 2 | – | 1 | – | – |
| **(8)** | **125** | **109** | **102** | **84** | **76** | **67** | **55** | **44** | **34** | **23** | **14** | **8** | **7** | **3** | **2** | **2** | **1** |

(b)

90th percentile suspended sediment concentrations ($mg \, l^{-1}$) at 2 km intervals upstream from the mouth

| Run | | | | | | | | | | | | | | | | | |
|-----|---|---|---|---|---|---|---|---|---|---|---|---|---|---|---|---|---|
| (1) | 290 | 250 | 229 | 182 | 166 | 144 | 118 | 86 | 63 | 40 | 24 | 16 | 16 | 10 | 9 | 7 | 9 |
| (2) | 807 | 783 | 799 | 742 | 703 | 615 | 534 | 420 | 336 | 206 | 147 | 111 | 98 | 72 | 63 | 55 | 45 |
| (3) | 279 | 259 | 257 | 218 | 208 | 186 | 171 | 148 | 114 | 81 | 63 | 48 | 52 | 31 | 24 | 22 | 18 |
| (4) | 277 | 249 | 237 | 188 | 177 | 165 | 141 | 115 | 77 | 44 | 24 | 17 | 15 | 8 | 6 | 4 | 3 |
| (5) | 90 | 84 | 82 | 70 | 67 | 62 | 53 | 44 | 37 | 28 | 18 | 16 | 16 | 8 | 6 | 7 | 5 |
| (6) | 388 | 347 | 328 | 265 | 254 | 240 | 210 | 151 | 90 | 55 | 33 | 22 | 17 | 8 | 7 | 5 | 1 |
| (7) | 149 | 137 | 133 | 107 | 99 | 93 | 81 | 59 | 36 | 18 | 11 | 8 | 7 | 4 | 4 | 1 | 1 |
| **(8)** | **278** | **250** | **245** | **199** | **183** | **163** | **148** | **121** | **82** | **46** | **29** | **21** | **19** | **10** | **7** | **8** | **5** |

(c)

| | Suspended | | Deposited | | Exchange | | Deposited per year | % deposit exchange | Average suspended |
|-----|------|--------|-------|--------|-------|---------|------|------|------|
| Run | Neap | Spring | Neap | Spring | Neap | Spring | | | |
| (1) | 8.58 | 70.21 | −0.88 | 10.73 | 9.14 | 182.69 | 2200 | 3.8 | 31.03 |
| (2) | 21.91 | 476.73 | −1.20 | 118.22 | 11.49 | 1020.43 | 9300 | 5.9 | 167.55 |
| (3) | 8.60 | 124.63 | −1.17 | 30.70 | 8.41 | 289.48 | 3200 | 5.1 | 47.21 |
| (4) | 7.78 | 121.18 | −0.99 | 26.61 | 7.72 | 285.81 | 2700 | 4.3 | 45.21 |
| (5) | 4.08 | 23.55 | −0.58 | 2.83 | 4.44 | 60.80 | 800 | 3.5 | 11.84 |
| (6) | 19.11 | 134.84 | −1.51 | 23.94 | 17.92 | 310.08 | 3000 | 3.6 | 62.73 |
| (7) | 6.62 | 61.39 | −0.48 | 6.31 | 5.28 | 134.45 | 800 | 2.6 | 26.10 |
| **(8)** | **8.32** | **116.75** | **−1.59** | **24.30** | **6.71** | **271.36** | **2500** | **4.2** | **44.54** |

*Note:* Units: $10^3$ tonnes
Parameters as in Table 8.2.
Diameter 22 μm, $W_s = 0.0005 \, m \, s^{-1}$.
*Source:* Lane and Prandle, 2006.

a corresponding deposition rate of 1 Mt per year. While capture rates increase progressively with increasing sediment size (above $d = 30 \, \mu m$), corresponding decreases in concentration yield a maximum deposition at 60 μm of 2 Mt per year. This maximum is close to the preponderance of sediments with $W_S = 0.003 \, m \, s^{-1}$

($d = 54\ \mu m$) found by Hill *et al.* (2003). In Section 7.5, it was shown that the size of suspended sediments corresponding to 'equilibrium' conditions of zero net deposition or erosion is in the range 20–50 $\mu m$. Throughout the range of $d = 30$–100 $\mu m$, net sedimentation remains surprisingly constant at between 1 and 2 Mt per year. This sedimentation rate is in close agreement with observational evidence (Table 8.1).

### 8.2.6 Sensitivity to model parameters

The model's responses to the following parameters were quantified: vertical structure of currents, eddy diffusivity, salinity, the bed friction coefficient and sediment supply.

Table 8.3 shows, for $W_S = 0.0005\ \mathrm{m\ s}^{-1}$ ($d = 22\ \mu m$), the sensitivity to Run numbers:

(**R1**) – No vertical current shear, i.e. a 2D hydrodynamic model.
(**R2**) – Depth-varying eddy diffusivity, $K_z(z) = K_z(-3z^2 + 2z + 1)$, i.e. depth-mean value $K_z$ at the bed, 1.33 $K_z$ at $z = Z/D = 0.33$ and 0 at the surface
(**R3**) – A time varying value of $K_z(t)$, with a quarter-diurnal variation of amplitude 0.25 $K_z$ producing a peak value 1 h after peak currents.
(**R4**) – Mean salinity-driven residual current profile (4.15) $U_z = g\,S_x D^3/K_z(-0.1667z^3 + 0.2687z^2 - 0.0373z - 0.0293)$, where the salinity gradient $S_x$ was specified over a 40 km axial length.
(**R5**) – Bed friction coefficient halved, $f = 0.5 \times 0.0158\,W_s^{1/4}$.
(**R6**) – Bed friction coefficient doubled, $f = 2.0 \times 0.0158\,W_s^{1/4}$.
(**R7**) – Erosion rate at the mouth $0.5\,\gamma$, i.e. halving the rate of supply of marine sediments.
(**R8**) – Baseline simulation.

While the calculated values of sediment concentration and net fluxes varied widely and irregularly, the net deposition remained much more constant. The acute and complex sensitivity to bed roughness and related levels of eddy diffusivity and viscosity is evident from Table 8.3. The acute sensitivity to bed roughness and sediment supply leads to concern that migration of new flora and fauna might lead to 'modal shifts' with potentially dramatic consequences.

To comprehend these sensitivities, in shallow water we can approximate, from Prandle (2004), the following dependencies on the friction factor $f$:

| | |
|---|---|
| Tidal velocity amplitude | $U^* \sim f^{-1/2}$ |
| Sediment concentration | $C \sim f^{1/2}$ |
| Tidal sediment flux | $U^*C \sim f^0$ |
| Residual sediment flux | $<UC> \sim U^*C \cos\theta \sim f^{1/2},$ |

where $\theta$ is the phase lag of tidal elevation relative to currents and residual sediment flux corresponds to net upstream deposition. These theoretical results are consistent with the increases in concentration and residual fluxes for larger values of $f$ shown by the model for both sediment types.

### 8.2.7 Summary

A century of bathymetric surveys indicate a net loss of estuarine volume of about 0.1%, or 1 million cubic metres, per year. Similar percentage losses are found in many of the large estuaries of NW Europe. Sea level rise of $1.2 \, \text{mm} \, \text{a}^{-1}$ represents only a 0.02% annual increase. This relative stability persists in a highly dynamic regime with suspended sediment concentrations exceeding $2000 \, \text{mg} \, \text{l}^{-1}$ and spring tide fluxes of order $200\,000 \, \text{t}$. Detailed analyses of the bathymetric sequences indicate that most significant changes occur in the upper estuary and in inter-tidal zones. Long-period, up to 63 years, tidal elevation records in the lower estuary show almost no changes to the predominant $M_2$ and $S_2$ constituents.

A 3D Eulerian fine-resolution hydrodynamic model coupled with a Lagrangian, random-walk sediment module was used to show how the dominant fluxes involve fine (silt) sediments on spring tides. The closest agreement between observed and model estimates of net imports of sediments occurs for sediments of diameter of approximately $50 \, \mu\text{m}$ – both dredging records and *in situ* observations indicate that sediments of this kind predominate. The model showed little influence of river flow, saline intrusion or channel deepening on the sediment regime. Conversely, the net fluxes were sensitive to both the bed friction coefficient, $f$, and the phase difference, $\theta$, between elevation tidal velocity and elevation.

Upper-bound rates of infill of up to $10 \, \text{Mt} \, \text{a}^{-1}$ are indicated by the model, comparable with annual dredging rates of up to $5 \, \text{Mt}$. The limited mobility of coarse sediments was contrasted with the near-continuously suspended nature of the finest clay. While the model indicated that sedimentation rates might increase significantly for much finer particles, this is likely to be restricted by the limited availability of such material in the adjacent coastal zone. The present approach can be readily extended to study changes in biological mediation of bottom sediments, impacts of waves, consolidation and the interactions between mixed sediments.

### 8.3 Impacts of GCC

By 2050, GCC could significantly change mean sea levels, storminess, river flows and, hence, sediment supply in estuaries (IPCC, 2001). The tidal and surge response within any estuary will be further modified by accompanying natural morphological (post-Holocene) adjustments alongside impacts from past and present 'interventions'. Generally, relatively small and gradual morphological adjustments are expected.

As an illustration, deposition per tide of a depth-mean concentration of $100 \, \mathrm{mg} \, \mathrm{l}^{-1}$ in $10 \, \mathrm{m}$ water depth amounts to about $0.35 \, \mathrm{mm}$, or $25 \, \mathrm{cm}$ per year. In reality, as shown in the Mersey, 'capture rates' (upstream deposition as a proportion of the net tidal inflow of suspended sediments) are typically only a few percent. Thus, simulations need to extend over decades to embrace responses over the full range of forcing cycles involved. However, as noted previously, longer-term extrapolations with 'Bottom-Up' models become increasingly chaotic, and hence, here we examine impacts from GCC by using the Theoretical Frameworks developed in earlier chapters.

### 8.3.1 Impacts on tide and surge heights

The response Framework, Fig. 2.5, based on analytical expressions derived by Prandle and Rahman (1980), provides immediate indications of likely changes in the estuarine response of tides and surges. Figure 2.5 shows that amplification of tides (and surges) between the first 'node' and the head of the estuary can be up to a factor of 2.5. Concern focuses on conditions in estuaries where, for the excitation 'period', $P$, the bathymetric dimensions (length, depth and shape) result in the estuarine mouth coinciding with this node with consequent resonant amplification. This occurs when, (2.26),

$$y = 0.75 \, v + 1.25, \tag{8.2}$$

$$\text{where } y = \frac{4 \pi L}{P(2 - m)(gD)^{1/2}} \text{ and } v = \frac{n + 1}{2 - m}.$$

$L$ and $D$ are the estuarine length and depth (at the mouth), and $m$ is the power of axial depth variation and $n$ of breadth variation.

The estuarine length, $L_R$, for maximum amplification is then

$$L_R = (2 - m)(0.75v + 1.25) \, g^{1/2} \, D^{1/2} \, \frac{P}{(4\pi)}. \tag{8.3}$$

The Framework extends from $0 < v < 5$ encompassing the following range of shapes and associated resonant lengths:

(a) canal        $m=n=0, \quad v=0.5 \quad L_c = 0.25(gD)^{1/2}P$
(b) embayment    $m=n=0.5, v=1 \quad \; L= 3/3.25L_c$
(c) linear       $m=n=1, \quad v=2 \quad \; L= 2.75/3.25L_c$          (8.4)
(d) funnel       $m=n=1.5, v=5 \quad \; L= 2.5/3.25L_c.$

Thus, as shown in Section 2.4.1, the range of funnelling (b) to (d) results in a relatively small reduction in the 'quarter-wavelength' resonant length applicable for a prismatic channel. Figure 8.5 indicates corresponding resonant periods for a

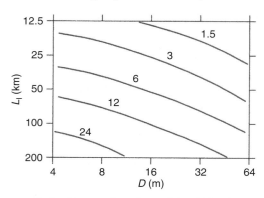

Fig. 8.5. Resonant periods (h) as a function of depth (at the mouth) and length. Results are for linear axial variations in depth and breadth, but from (8.4) are more widely applicable.

semi-diurnal tidal period for a 'linear' estuary with $m = n = 1$. From (8.4), these results are broadly applicable for a wide range of estuarine shapes. The figure shows that even for a depth at the mouth of 4 m, resonance at semi-diurnal frequencies will only occur for $L_R > 60$ km, while for $D = 16$ m, $L_R > 100$ km. From (8.3), for a synchronous estuary ($m = n = 0.8$, $v = 1.5$), $L_R = 37\ D^{1/2}$ (km) or 94 km for the mean observed depth, $D = 6.5$ m. This emphasises that only the longest of UK estuaries, such as the Bristol Channel, are likely to exhibit significant tidal amplification.

Using the expression (6.12) for the length of a synchronous estuary

$$L = \frac{120\ D^{5/4}}{(f\varsigma^*)^{1/2}} \tag{8.5}$$

with the bed friction coefficient $f = 0.0025$ and $\varsigma^*$ tidal elevation amplitude. By inserting (8.3) into (8.5), we derive the following expressions for resonant values of $L_R$ and $D_R$ in terms of tidal amplitude, $\varsigma^*$:

$$L_R(\text{km}) \sim 180\ \varsigma^{*1/3} \tag{8.6}$$

and

$$D_R(\text{m}) \sim 31\varsigma^{*2/3}. \tag{8.7}$$

Hence, for $\varsigma^* = 1$ m, $L_R = 180$ km and $D_R = 31$ m while for $\varsigma^* = 4$ m, $L_R = 285$ km and $D_R = 78$ m.

Thus, we only anticipate resonance at the semi-diurnal frequency in deep systems such as the Bristol Channel where the estuarine 'resonance' extends to the adjacent shelf sea. Hence, we do not expect dramatic changes in tidal or surge responses in estuaries for anticipated changes in sea level of up to 1 m. Thus, increases in flood levels due to rises in msl are likely to be of the same order as the respective increases in adjacent open-sea conditions. Exceptions to the above are possible for surge response to secondary depressions which can have effective periodicities of significantly less than 12 h and hence corresponding reductions in 'resonant' estuarine lengths.

### 8.3.2 Bathymetric adjustments

Chapter 6 shows how, for 'synchronous' estuaries, a 'zone of estuarine bathymetry' can be determined bounded by

$$\frac{E_x}{L} < 1, \frac{L_1}{L} < 1 \text{ and } \frac{D}{U^3} < 50\,\text{m}^{-2}\text{s}^3, \tag{8.8}$$

corresponding to both tidal excursion, $E_x$, and salinity intrusion length, $L_I$, being less than estuarine length $L$ and the Simpson–Hunter (1974) criterion, $D/U^3$, for a 'mixed' estuary.

Figure 6.12 shows how Bar-Built and Coastal Plain estuaries in the UK generally fit within this bathymetric zone. By introducing the expression (6.25) linking depth at the mouth to river flow and side slope gradient $\tan\alpha$

$$D_0 = 12.8(Q\tan\alpha)^{0.4}. \tag{8.9}$$

Figures 7.9 and 7.10 show comparisons between observed lengths and depths (at the mouth) against the theoretical values (8.5) and (8.9). The observed values were extracted by Prandle *et al.* (2006) from the 'FutureCoast' database (Burgess *et al.*, 2002).

These Frameworks, Figs 6.12, 7.9 and 7.10, then provide immediate visual indications of the likely stability and sensitivity of any particular estuary to changes in $D$, $Q$ or $\varsigma^*$. Estuaries located within the bathymetric zone in Fig. 6.12, with depths and lengths in broad agreement with the theoretical values in Figs 7.9 and 7.10, might be regarded as in present-day dynamic equilibrium. Consequently, future morphological adjustments might be expected to remain consistent with these theories and follow relatively rapidly. By contrast, estuaries outside the zone or where either depth or length is inconsistent with the theories might suggest anomalous characteristics. By identifying the bases of such anomalies, the implications for future morphology can be assessed.

Clearly a sea level rise, of say 1 m, will have a much bigger impact on shallow estuaries than deep. Prandle (1989) examined the change in tidal response in estuaries due to variations in msl, where the locations of the coastal boundaries remained fixed (i.e. construction of flood protection walls). The results showed the largest impacts in long, shallow estuaries.

The theories synthesised in Figs 6.12, 7.9 and 7.10 do not consider sedimentation. Changes in the nature and supply of marine sediments can lead to abrupt changes in estuarine morphology. This supply can directly determine the nature of the surficial sediments and thereby bed roughness. Changing flora and fauna can, via their effects on sea-bed roughness and associated erosion and deposition rates, have abrupt and substantial impacts on dynamics and bathymetry. Peculiarly, the relationship (8.9) between depth at the mouth and river flow is independent of both tidal amplitude and bed roughness. However, from (8.5), the associated estuarine length will shorten as sediments become coarser.

### 8.3.3 Depth, breadth and length changes for 2100 variations in msl and river flow

Estimates of 'precautionary' changes in msl by 2100 (Defra/Environment Agency Technical Summaries, 2003 and 2004) amount to an increase of 50 cm. Corresponding estimates for river fluxes include both increases and decreases of up to 25%.

Inserting these changes in river flow, $Q$, into (8.9) and the resulting changes in depth, $\delta D$, into (8.5), we can estimate the changes in length, $\delta L$. Likewise, the changes in breadth, $\delta B$, associated with the changes in $D$ can be estimated by assuming the side-slope gradients, $\tan \alpha$, are unchanged. Table 8.4 provides quantitative indications of the resultant changes. The representative values of $D$, $L$ and $B$ over the range of estuarine geomorphologies were calculated from the FutureCoast data set (Prandle, 2006).

The changes $\delta D$ correspond to $\delta Q^{0.4}$, changes $\delta L$ to $(\delta Q^{0.4})^{1.25}$ and $\delta B$ to $2\delta D / \tan \alpha$. The results show that, on average, the 'dynamical' adjustment to a 25% change in river flows may change depths by as much as the projected sea level

Table 8.4 *Changes in depth, length and breadth for a 25% change in river flow and 0.5 m increase in msl*

| Estuary type | $D$ (m) | $\delta D_Q$ ($\pm$) | $L$ (km) | $\delta L_Q$ ($\pm$) | $\delta L_{msl}$ (+) | $B$ (m) | $\delta B_Q$ ($\pm$) | $\delta B_{msl}$ (+) |
|---|---|---|---|---|---|---|---|---|
| All minimum | 2.5 | 0.25 | 5 | 0.62 | 1.28 | 130 | 38 | 77 |
| Mean | 6.5 | 0.65 | 20 | 2.50 | 1.94 | 970 | 100 | 77 |
| All maximum | 17.3 | 1.73 | 41 | 5.12 | 1.49 | 3800 | 266 | 77 |
| Coastal Plain | 8.1 | 0.81 | 33 | 4.12 | 2.57 | 1500 | 147 | 91 |
| Bar-Built | 3.6 | 0.36 | 9 | 1.12 | 1.59 | 510 | 51 | 71 |

*Notes:* Change in river flow – subscript '$Q$', 0.5 m increase in msl – subscript 'msl'.
*Source:* Prandle, 2006.

rise – with this effect reduced in smaller estuaries and significantly increased in larger ones. The resulting changes in estuarine lengths and breadths follow similar patterns with the bigger 'dynamical' changes occurring in the larger estuaries where they are significantly greater than those due to the specified sea level rise. Overall, we anticipate changes in estuarine lengths of the order of 0.5–5 km and breadths of the order 50–250 m due to the 25% change in river flow. Corresponding changes due a sea level rise of the order 50 cm involve increases in both lengths of order 1–2.5 km and breadths of order 70–100 m.

### 8.3.4 Impacts on currents, stratification, salinity, flushing and sediments

Indicative impacts of GCC on the above parameters can be similarly calculated using the respective parameter dependencies and Theoretical Frameworks summarised in Section 1.5. While the peculiar conditions in any specific estuary will determine the actual response, such immediate indications can provide useful perspectives.

## 8.4 Strategies for modelling, observations and monitoring

### 8.4.1 Modelling

Coupled hydrodynamic and mixing models are required as the basis for transporting and mixing contaminants both horizontally and vertically. The dynamical processes involved occur over time scales of seconds (turbulent motions), to hours (tidal oscillations), to months (seasonal variations) with corresponding space scales from millimetres to kilometres. In addition to these hydrodynamic and mixing models, sediment and ecological models are required with robust algorithms for sources, sinks and biological/chemical reactive exchanges for longer-term simulations.

Both proprietary and public-domain model codes typically involve investment of tens of years in software development and continued maintenance by sizeable teams. Such effort is increasingly beyond the resources of most modelling groups, and standardised, generic modules in readily available public-domain codes are likely to be widely adopted. The development of such modules has removed much of the mystique that traditionally surrounded modelling of marine processes. The diversity of estuaries makes it unlikely that a single integrated model will evolve. Moreover, retention of flexibility at the module level is both necessary and desirable to accommodate a wide range of applications and to provide ensemble forecasts. Further developments of Theoretical Frameworks are important to interpret such ensemble simulations.

To understand and quantify the full range of threats from GCC, whole-system models are required – incorporating the impacts on marine biota and their potential biogeographic consequences. The introduction of various 'Water Framework Directives' for governance of regional seas and coasts emphasises the need for development of well-validated, reliable models for simulating water quality, ecology and, ultimately, fisheries. A systems approach is needed, capable of integrating marine modules and linking these into holistic simulators (geological, socio-economic, etc.). Rationalisation of modules to ensure consistency with the latter is an important goal, together with standardisation of prescribed inputs such as bathymetry and tidal boundary conditions. Such enhanced rationalisation will enable the essential characteristics of various types of models to be elucidated including the inherent limits to predictability.

In practice, coupling might be limited to sub-set representations (statistical emulators) encapsulating integrated parameters such as stratification levels or flushing times. To overcome the limitations of individual modules in such total-system simulations, methodologies are required both to quantify and to incorporate the range of uncertainties associated with model set-up, parameterisation and (future scenario) forcing. This requirement can be achieved by ensemble simulations providing relative probabilities of various outcomes linked to specific estimates of risk.

Model simulations and assessments should extend beyond a single ebb and flood cycle to include the spring–neap tidal cycle and seasonal variations in river flows and related density structures. Clearer insights and understanding of scaling issues should emerge by comparing modelling results with the new Theoretical Frameworks and against as wide a range of observational data as can be obtained.

### 8.4.2 Observations

Successful applications of models are generally limited by the paucity of resolution in observational data (especially bathymetry) used for setting-up, initialising, forcing (meteorological and along model boundaries), assimilation and validation. This paucity of data is a critical constraint in environmental applications. More and better observational data, extending over longer periods, are essential if modelling accuracy and capabilities are to be enhanced.

Instrumentation is lagging seriously behind model development and application, and this gap is expected to widen. A new generation of instrumentation is needed for the validation of species-resolving ecosystem models. Despite recent advances, the range of marine parameters that can be accurately measured is severely restricted and the cost of observations is orders of magnitude greater than that associated with models.

Comprehensive observational networks are needed exploiting synergistic aspects of the complete range of instruments and platforms and integrally linked to modelling requirements. Permanent *in situ* monitoring is likely to be the most expensive component of any observational network, and it is important to optimise such networks in relation to the modelling system for the requisite forecasts.

To define estuarine boundary conditions, there is a related requirement for accurate (model) descriptions of the state of adjacent shelf seas. Permanent coastal monitoring networks have been established in coastal seas and estuaries measuring water levels, currents profiles, surface winds, waves, temperature, SPM, salinity, nutrients, etc. using tide gauges, mooring and drifting buoys, platforms, ferries alongside remote sensing from satellites, radar and aircraft. Regional monitoring networks are being established via the Global Oceanographic Observing System (GOOS) networks, (UNESCO, 2003).

Up-scaling of knowledge from small-scale experimental measurements is required to provide larger- and longer-term algorithms employed in numerical models. Test-bed observational programmes are needed to assess model developments, these should ideally extend to water levels, currents, temperature and salinity, waves, turbulence, bed features, sedimentary, botanical, biological and chemical constituents. To maximise the value of such observations to the wider community, results should be made available in complete, consistent, documented and accessible formats.

### *8.4.3 Monitoring*

A basic monitoring strategy for studying bathymetric changes, capable of better resolving processes operating in estuaries, should include the following:

(1) shore-based tide gauges throughout the length of an estuary, supplemented by water level recorders in the deeper channels;
(2) regular bathymetric surveys, e.g. 10-year intervals with more frequent re-surveying in regions of the estuary where low water channels are mobile;
(3) a network of moored platforms with instruments for measuring currents, waves, sediment concentrations, temperature and salinity.

Maximum use should be made of the synergy between satellite, aircraft, ship, sea surface, seabed and coastal (radar) instrumentation (Prandle and Flemming, 1998). Likewise, new assimilation techniques should be used for bridging gaps in monitoring capabilities. Observer systems sensitivity experiments can be used to determine the value of the existence or omission of specific components in a new or existing monitoring system.

## 8.5 Summary of results and guidelines for application

Strategic planning to address long-term sustainability of estuaries needs to make full use of developments in modelling, monitoring and theory. New Theoretical Frameworks provide a perspective on the threat from GCC.

The leading question is:
***How will estuaries adapt to GCC?***

### 8.5.1 Challenges

Management challenges include

(1) promoting sustainable exploitation, i.e. permitting commercial and industrial development subject to assessment of associated impacts, e.g. dredging, reclamation and fish-farming
(2) satisfying national and international legislation and protocols relating to discharges;
(3) improving and promoting the marine environment, monitoring water quality, supporting diverse habitats and expanding recreational facilities;
(4) reducing risks in relation to flooding, navigation and industrial accidents;
(5) long-term strategic planning to address future trends including GCC.

A major difficulty in estuarine management is the general uncertainty in linking specific actions to subsequent responses over the local to wider scales and from the immediate to longer time scales. For example, it has generally proved difficult to predict improvements to estuarine water quality following clean-up campaigns due to leaching of contaminants from historical residues in bed sediments. Similarly, the full impacts from 'interventions' may manifest themselves in unforeseen ways at remote sites at a much later time. While such uncertainties can never be entirely overcome, the pragmatic objective is to arrive at a balanced perspective. This perspective should provide indications of the scale of vulnerability based on an ensemble of 'approaches' using theory, measurements and modelling to draw on present and past behaviour of the estuary concerned and on related experiences in adjacent systems and in similar estuaries elsewhere.

### 8.5.2 Modelling case study

Section 8.2 describes a case study of a modelling simulation of the Mersey Estuary, indicating how theory and observational data are used to assess the model results and interpret parameter sensitivity tests. Figures 8.2 and Table 8.1 illustrate the use of results from earlier modelling and observational studies. Figures 8.3 and 8.4 show results from a random-walk particle representation of sediment transport. This study highlights the value of long-term observational data sets. Unfortunately, such data sets are overwhelmingly from large (navigable) estuaries and, as such, can be misleading in relation to experience in smaller, shallower estuaries. Thus, in Section 8.3, the acute sensitivity of 'near-resonant' tidal responses in the Bay of Fundy (Garrett, 1972) and the Bristol Channel (Prandle, 1980) are shown to be exceptional. Figure 8.5 indicates that resonant response for semi-diurnal tidal constituents only occurs in estuaries greater than 60 km in length. The related demarcation between 'inertially dominated' systems and the much more commonly encountered, shorter and shallower 'frictionally dominated' systems was shown in Fig. 6.3.

### 8.5.3  Strategic planning

Section 8.4 considers future modelling and observational strategies. Strategic planning for estuarine sustainability must encompass the wide spectra of temporal, spatial and parameter scales encompassing physics to ecology and micro-turbulence to whole estuary circulation. Advantage must be taken of the rapid advances in numerical modelling with the associated growth in computational power, monitoring technologies and scientific understanding. However, securing investment in such advances generally requires demonstrable benefits to end users.

Faced with specific planning issues such as a proposed engineering 'intervention' or the need to improve flood protection, managers will often commission a modelling study. The range of models available and their requirements was outlined in Section 1.4. It was indicated that the selection of an appropriate model depends on the availability of observational data to set-up, initialise, force, validate and assess simulations. Obtaining data is almost always much more expensive than a model study.

It is important to distinguish between model studies which involve 'interpolation' as opposed to 'extrapolation'. The results illustrated in Tables 8.2–8.4 are essentially 'interpolation', i.e. examining small perturbations close to existing parameter ranges. By contrast, 'extrapolation' involves larger perturbations which can change the ranking of controlling processes and introduce new elements outside of the range of validity of the model.

Ideally, an estuarine manager should have access to a range of modelling capabilities, routinely assessed by a wide range of continuously monitored data. Confidence in future predictions then rests on the degree to which such modelling systems can reproduce observed cycles, patterns and trends and interpret these against Theoretical Frameworks. Development of a strategic programme needs to exploit all such technologies to provide the robust perspective required both for long-term strategic planning and addressing specific day-to-day issues.

### 8.5.4 *Impacts of GCC*

The success of new theories in explaining the evolution of morphologies over the past 10 000 years of Holocene adjustments lends confidence for their use in extrapolation over the next few decades. The explicit analytical formulae and Theoretical Frameworks, summarised in Section 1.5, can provide guidance on the relative sensitivity of an estuary to both local 'intervention' and wider-scale impacts such as GCC. Figures 6.12, 7.9 and 7.10 represent new morphological frameworks. For any particular estuary, examining the loci on such frameworks from mouth to head and over the range of prevailing conditions can provide a perspective on the relative stability. Where these loci extend outside of the theoretical zones, the possibility of anomalous responses can be anticipated.

We do not expect drastic changes in estuarine responses to tides or surges from the projected impacts of GCC over the next few decades. Some enhanced sensitivity might be found in relation to shorter 'period' (6 h) surges associated with secondary depressions, particularly in larger estuaries. Maintaining fixed boundaries in the face of continuous increases in msl may enhance surge response in the shallowest estuaries (Prandle, 1989).

In the absence of 'hard geology', enhanced river flows may result in small increases in estuarine lengths and depths, developing over decades. By 2100, we anticipate changes in UK estuaries due to (precautionary) projected 25% changes in river flow: of order 0.5–5 km in lengths and of order 50–250 m in breadths. Corresponding changes due to a projected sea level rise of 50 cm are increases in lengths of order 1–2.5 km and breadths of order 70–100 m. In both cases, bigger changes will occur in larger estuaries. Although we do not expect dramatic impacts on sediment regimes, changing flora and fauna could, through their effect on sea-bed roughness and associated erosion and deposition rates, have abrupt and substantial impacts on dynamics and bathymetry.

Ultimately, an international approach is necessary to quantify the contribution to and effect from GCC. This extends to development of models and instruments (and their platforms), planning of monitoring strategies, exchange of data, etc. The pace of progress will depend on successful collaboration in developing structured research, development and evaluation programmes. The ultimate goal is a fusion of environmental data and knowledge, utilising fully the continuous development of communications and computational capacities. Appendix 8A indicates technologies likely to be widely available to estuarine managers in the next decade or so.

### Appendix 8A

### 8A.1 *Operational oceanography*

Operational oceanography is defined as the activity of routinely making, disseminating and interpreting measurements of coasts, seas, oceans and the atmosphere to provide forecasts, nowcasts and hindcasts.

## *8A.2 Forecasting, nowcasting, hindcasting and assimilation*

### *Forecasting*

Forecasting includes real-time numerical prediction of processes such as storm surges, wave spectra and sea ice occurrence. Forecasts on a climatic or statistical basis may extend forward for hours days, months, years or even decades. Accumulation of errors, both from model inaccuracies and from uncertainties in forcing, limits realistic future extrapolations.

### *Nowcasting*

In nowcasting, observations are assimilated in numerical models and the results are used to create the best estimates of fields at the present time, without forecasting. These observations may involve daily or monthly descriptions of sea ice, sea surface temperature, toxic algal blooms, state of stratification, depth of the mixed layer or wind–wave data.

### *Hindcasting*

Observational data for hindcasting are assimilated into models to compile sets of historic fields and distributions (typically monthly or annually) of variables such as sea surface elevation, water temperature, salinity, nutrients, radio-nuclides, metals and fish stock assessments.

### *Assimilation*

Data assimilation forms the interface between models, observation and theory and, thus, is an essential component in simulation systems (Fig. 8A.1). Assimilation is used to transfer observed information to update the model state, the model forcing and/or the model coefficients. The challenge is to take advantage of the complementary character of models and observations, i.e. the generic, dynamically continuous character of process knowledge embedded in models alongside the specific character of observed data.

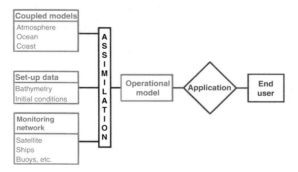

Fig. 8A.1. Components of an operational modelling system.

### 8A.3  Model generations

Numerical modelling has been used in marine science for almost 50 years. A convenient distinction is as follows:

*Generation 1*:    *exploratory models* where algorithms, numerical grids and schemes are being developed often utilising specific measurements focused on process studies.

*Generation 2*:    *pre-operational models* with (effectively) fully developed codes undergoing appraisal and development, generally against temporary observational measurements or test-bed data sets.

*Generation 3*:    *operational models* in routine use and generally supported by a permanent monitoring network, such as that shown in Fig. 8A.2.

A cascade time of approximately 10 years is typically required to migrate between each generation.

Real-time *operational* uses include tidal predictions and hazard warning for storm surges, oil or chemical spill movement, search and rescue, eutrophication, toxic algal blooms, etc. *Pre-operational* simulations often involve assessing and understanding the health of marine ecosystems and resources and their likely sensitivity to changing

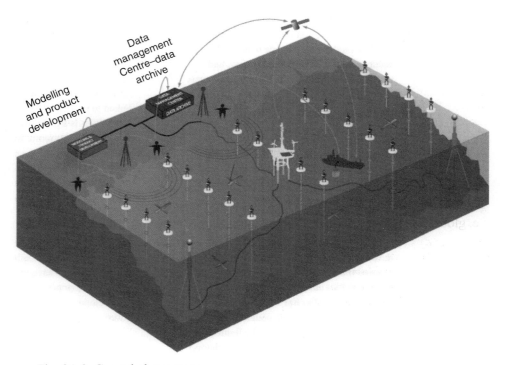

Fig. 8A.2. Coastal observatory.

conditions. These are typically concerned with assessment of absorptive capacity for licensing of discharges, evaluating environmental impacts of intervention (reclamation, dredging, etc.) and climate change. *Exploratory* applications extend from formulation of environmental management policies to developing the underpinning science and technology to address both anthropogenic influences and natural trends.

### *8A.4 Forecasting*

Even though the immediate issues of concern may seem far-removed from real-time 'operational' forecasting, recognition of global-scale developments in modelling of marine systems is important. Ultimately, estuarine research needs to be linked to the parallel progress in operational oceanography on both regional and global scales directed by GOOS (UNESCO, 2003). Effective operation of real-time forecasts requires the resources of a meteorological agency for communications, processing and dissemination of forcing data, alongside oceanographic data centres responsible for dissemination of quality-controlled marine data.

A major objective of operational oceanography is to minimise damage from future events by reducing uncertainties in forecasting, ranging from storms on a short time scale to rising sea levels and temperatures over the longer term. Operational oceanography is central to sustainable exploitation and management of our marine resources.

Progress in aircraft and satellite remote sensing (Johannessen *et al.*, 2000) will dictate the rate of development of operational oceanography for many parameters. Lead times of a decade or more are required for development of new sensors, commercial production of prototype instruments and international agreement on new satellite programmes. Remotely sensed data must be processed in hours if it is to be useful in operational forecasting. The need for enhanced information from atmospheric models is a high-priority item in operational forecasting. As an example, accuracy and extent, in time ahead, of wind forecasts are the primary limiting factors for wave and surge forecasting. The ultimate goal is dynamical coupling of estuaries–seas–ocean marine models through to terrestial and atmospheric modules, i.e. global integration of water, thermal and chemical budgets (Prandle *et al.*, 2005).

### References

Burgess, K.A., Balson, P., Dyer, K.R., Orford, J., and Townend, I.H., 2002. FutureCoast – The integration of knowledge to assess future coastal evolution at a national scale. In: *The 28th International Conference on Coastal Engineering. American Society of Civil Engineering*, Vol. 3. Cardiff, UK, New York, pp. 3221–3233.

Defra/Environment Agency, 2003. *Climate Change Scenarios UKCIP02: Implementation for Flood and Coastal Defence*. R&D Technical Summary W5B-029/TS.

Defra/Environment Agency, 2004. *Impact of Climate Change on Flood Flows in River Catchments*. Technical Summary W5-032/TS.

Fischer, H.B., List, E.J., Koh, R.C.Y., Imberger, J., and Brooks, N.H., 1979. *Mixing in Inland and Coastal Waters*. Academic Press, New York.

Garrett, C., 1972. Tidal resonance in the Bay of Fundy. *Nature*, **238**, 441–443.

Hill, D.C., Jones, S.E., and Prandle, D., 2003. Dervivation of sediment resuspension rates from acoustic backscatter time-series in tidal waters. *Continental Shelf Research*, **23** (1), 19–40, doi:10.1016/S0278-4343(02) 00170-X.

Hutchinson, S.M. and Prandle, D., 1994. Siltation in the saltmarsh of the Dee Estuary derived from $^{137}$Cs analysis of shallow cores. *Estuarine, Coastal and Shelf Science*, **38** (5), 471–478.

IPCC, 2001. Edited by Watson, R.T. and Core Writing Team. *Climate Change 2001: Synthesis Report. A Contribution of Working Groups I, II, and III to the Third Assessment Report of the Intergovernmental Panel on Climate Change*. Cambridge University Press, Cambridge, United Kingdom, and New York.

Johannessen, O.M., Sandven, S., Jenkins, A.D., Durand, D., Petterson, L.H., Espedal, H., Evensen, G., and Hamre, T., 2000. Satellite earth observations in Operational Oceanography. *Coastal Engineering*, **41** (1–3), 125–154.

Lane, A., 2004. Bathymetric evolution of the Mersey Estuary, UK, 1906–1997: causes and effects. *Estuarine, Coastal and Shelf Science*, **59** (2), 249–263.

Lane, A. and Prandle, D., 2006. Random-walk particle modelling for estimating bathymetric evolution of an estuary. *Estuarine, Coastal and Shelf Science*, **68** (1–2), 175–187, doi:10.1016/j.ecss.2006.01.016.

Pethick, J.S., 1984. *An Introduction to Coastal Geomorphology*. Arnold, London.

Prandle, D., 1980. Modelling of tidal barrier schemes: an analysis of the open-boundary problem by reference to AC circuit theory. *Estuarine and Coastal Marine Science*, **11**, 53–71.

Prandle, D., 1989. *The Impact of Mean Sea Level Change on Estuarine Dynamics*. C7-C14 in Hydraulics and the environment, Technical Section C: Maritime Hydraulics. Proceedings of the 23rd Congress of the IAHR, Ottawa, Canada.

Prandle, D., 2004. How tides and river flows determine estuarine bathymetries. *Progress in Oceanography*, **61**, 1–26, doi:10.1016/j.pocean.2004.03.001.

Prandle, D., 2006. Dynamical controls on estuarine bathymetries: assessment against UK database. *Estuarine, Coastal and Shelf Science*, **68** (1–2), 282–288, doi:10.1016/j. ecss.2006.02.009.

Prandle, D. and Flemming, N.C. (eds.), 1998. *The Science Base of EuroGOOS. EuroGOOS*, Publication, No. 6. Southampton Oceanography Centre, Southampton.

Prandle, D. and Rahman, M., 1980. Tidal response in estuaries. *Journal of Physical Oceanography*, **10** (10), 1522–1573.

Prandle, D., Lane, A., and Manning, A.J., 2006, New typologies for estuarine morphology. *Geomorphology*, **81** (3–4), 309–315.

Prandle, D., Los, H., Pohlmann, T., de Roeck Y-H., and Stipa, T., 2005. *Modelling in Coastal and Shelf Seas – European Challenge. ESF Marine Board Postion Paper 7*. European Science Foundation, Marine Board.

Prandle, D., Murray, A., and Johnson, R., 1990. Analyses of flux measurements in the River Mersey. pp 413–430 In: Cheng, R.T. (ed.), *Residual Currents and Long Term Transport, Coastal and Estuarine Studies*, Vol. 38. Springer-Verlag, New York.

Price, W.A. and Kendrick, M.P., 1963. Field and model investigation into the reasons for siltation in the Mersey Estuary. *Proceedings of the Institute of Civil Engineers*, **24**, 473–517.

Simpson, J.H. and Hunter, J.R., 1974. Fronts in the Irish Sea. *Nature*, **250**, 404–406.

Thomas, C.G., Spearman, J.R., and Turnbull, M.J., 2002. Historical morphological change in the Mersey Estuary. *Continental Shelf Research*, **22** (11–13), 1775–1794, doi:10.1016/S0278-4343(02)00037-7.

UNESCO, 2003. *The Integrated Strategic Design Plan for the Coastal Ocean Observation Module of the Global Ocean Observation System*. GOOS Report No. 125. IOC Information Documents Series No. 1183.

Woodworth, P.L., Tsimplis, M.N., Flather, R.A., and Shennan, I., 1999. A review of the trends observed in British Isles mean sea level data measured by tide gauges. *Geophysical Journal International*, **136** (3), 651–670.

# Index